"*Bayesian Statistics: The Basics* provides extremely clear and accessible language to the novice, guiding the reader through Bayesian statistics with care, via hands-on examples involving user-friendly open-source software. Whether you are a graduate student, researcher, or faculty member, you now have no excuse for not deeply understanding, appreciating, and applying the fundamentals of Bayesian statistics."

Fred Oswald, *Professor of Psychological Sciences, Rice University, USA*

"The book is written in a highly accessible manner, and what is most important to me is that it emphasizes the conceptual understanding behind Bayesian methods, which can be grasped even without a thorough understanding of underlying mechanics. I greatly appreciate the idea of a concise statistics book whose purpose is to build a solid conceptual framework that gives the reader a bird's-eye view of the subject. I think *Bayesian Statistics: The Basics* serves this purpose brilliantly."

Krzysztof Cipora, *Senior Lecturer in Mathematical Cognition, Loughborough University, UK*

"This is a highly accessible introduction to Bayesian statistics using JASP, a free, open-source program. The book features clear, concise explanations for beginners but will also provide new insights for people who already have a grasp of the basics. I particularly like that there are suggestions for how to report the results."

Mark LaCour, *Assistant Professor of Psychology, University of Louisiana at Lafayette, USA*

W0113178

"The book is an excellent introduction of both Bayesian statistics and JASP. Going through the examples in each chapter helped me understand both the theory and application simultaneously. I appreciated the stripped-down approach to covering the material with a very light focus on the underlying math. I now feel comfortable providing Bayesian results along with the classical results when I conduct experiments with these kinds of designs (e.g., needing correlation, t-test, ANOVA, or regression). Overall, I really enjoyed reading this book and it taught me a lot. I'm sure it will be popular and useful to many students, researchers, and faculty!"

Curt Carlson, *Professor of Psychology,*
East Texas A&M University, USA

"*Bayesian Statistics: The Basics* is written in a clear, digestible tone that turns 'daunting' into 'doable.' This tone is then supported with a scaffolding approach, clear definitions, and comprehensive steps that takes Bayesian statistics from 'doable' to 'enjoyable.' This textbook is the reference I wish I had as a student, and the tool I can't wait to use in the classroom."

Bryanna Scheuler, *PhD student in Psychology,*
The University of Texas at San Antonio, USA

"Dr Faulkenberry's writing style is so understandable and is great for students and researchers who are learning Bayesian statistics for the first time. One of my favorite parts of this book is the sample write-ups that are included at the end of each section. This ensures that researchers new to Bayesian statistics know how to properly communicate their Bayesian work in future publications. Using them as a model will help to make a Bayesian beginner write their statistics like a pro!"

Amy Bohmann, *Associate Professor of Psychology,*
Texas A&M University – San Antonio, USA

BAYESIAN STATISTICS

Bayesian Statistics: The Basics provides a comprehensive yet accessible introduction to Bayesian statistics, specifically tailored for any researcher with an interest in statistical methods. It covers the theoretical foundations of Bayesian inference, contrasting it with classical statistical methods like null hypothesis significance testing. The book emphasizes key concepts such as prior and posterior distributions, Bayes' theorem, and the Bayes factor, making them understandable even for readers with minimal mathematical backgrounds.

Methodologically, the book offers practical, step-by-step guides on how to conduct Bayesian analyses using the free software package JASP. Each chapter focuses on applying Bayesian methods to common research designs with real-world data. Readers will benefit from the clear examples, visualizations, and JASP screenshots that ensure the learning experience is interactive and easy to follow.

Full of practical content, the book emphasizes the advantages of Bayesian model comparison over traditional approaches, especially in quantifying evidence for competing hypotheses. Readers will also learn how to perform sensitivity analyses to assess the impact of different prior assumptions on their results.

By the end of the book, readers will get both the theoretical understanding and practical skills to implement Bayesian methods in their own research, making it an invaluable resource for both novice and experienced researchers studying Bayesian statistics.

Thomas J. Faulkenberry, PhD, is a professor of psychological sciences and associate dean of the College of Graduate Studies at Tarleton State University in Stephenville, TX (USA). A mathematician by training, he teaches courses on statistics and mathematical modeling in the behavioral sciences, and his primary research areas are mathematical cognition and Bayesian statistics.

The Basics Series

The Basics is a highly successful series of accessible guidebooks which provide an overview of the fundamental principles of a subject area in a jargon-free and undaunting format.

Intended for students approaching a subject for the first time, the books both introduce the essentials of a subject and provide an ideal springboard for further study. With over 50 titles spanning subjects from artificial intelligence (AI) to women's studies, *The Basics* are an ideal starting point for students seeking to understand a subject area.

Each text comes with recommendations for further study and gradually introduces the complexities and nuances within a subject.

MODERN ARCHITECTURE
GRAHAM LIVESEY

MICROECONOMICS
THOMAS R. SADLER

STATISTICAL ANALYSIS
CHRISTER THRANE

ANTHROPOLOGY OF REPRODUCTION
SALLIE HAN AND CECÍLIA TOMORI

SCIENCE COMMUNICATION
MASSIMIANO BUCCHI AND BRIAN TRENCH

PROJECTION DESIGN
DAVIN E. GADDY

ETHNOGRAPHY
SUSAN WARDELL

FAT STUDIES
MAY FRIEDMAN

BAYESIAN STATISTICS
THOMAS J. FAULKENBERRY

For more information about this series, please visit: www.routledge.com/The–Basics/book–series/B

BAYESIAN STATISTICS

THE BASICS

Thomas J. Faulkenberry

Routledge
Taylor & Francis Group

NEW YORK AND LONDON

Designed cover image: Getty Images via piranka

First published 2025
by Routledge
605 Third Avenue, New York, NY 10158

and by Routledge
4 Park Square, Milton Park, Abingdon, Oxon, OX14 4RN

Routledge is an imprint of the Taylor & Francis Group, an informa business

ISBN: 9781032746388 (hbk)
ISBN: 9781032744001 (pbk)
ISBN: 9781003470199 (ebk)

DOI: 10.4324/9781003470199

Typeset in Bembo
by Apex CoVantage, LLC

CONTENTS

PREFACE

By starting to read this book, I can confidently conclude that you already have some motivation to read about Bayesian statistics. In fact, I imagine that several people will pick up this book with some hope of entering the (perhaps mysterious) world of Bayesian statistics. However, I imagine that the path each person has taken to arrive at this motivation may be quite different from the others. Maybe you are a seasoned practitioner of classical statistical methods and you want to expand your toolbox to include Bayesian statistics. Maybe you are a graduate student in a field that is moving increasingly toward Bayesian methods, and you feel like you need a good starting place to learn. Maybe you already know quite a bit about Bayesian statistics, but you're looking for a short guidebook to recommend to your students and/or colleagues.

Regardless of the path you've taken to read these opening paragraphs, I want you to feel assured of one thing – this book is the right one for *you*.

Another thing is worth keeping in mind as you contemplate moving forward with this book – *all paths to learning are valid ones*. No matter what motivation drives you to join this journey with me, it is the right one for *you*.

It is possible that you've already taken a look at some of the other books about Bayesian statistics that are out there. Indeed, these books are included in my shelves as well, and I glean something new from them every time I open one. Certainly, Bayesian statistics is becoming an increasingly popular method for doing inference in a wide array of fields. Unfortunately, there are relatively few accessible textbooks that cover Bayesian statistics, and, of those that do,

most are either dense, expensive, too broad, mathematically impen-
etrable, or some combination of these descriptors. You may have
even bought a book or two, opened them with lots of hope, and
then closed them pretty quickly – not because any of them are bad
books, but because one's first foray into a subject needs a friendly
guide to show them the way.

Like fishing and bicycle touring, the most successful outings for
the beginner are those that are accompanied by a guide. Bayesian
statistics is no different. I hope this book will be your guide.

My goal in writing this book is to provide you with an inexpen-
sive, focused, clear, and accessible introductory book on Bayesian
statistics that will give you a chance to learn Bayesian methods in
a context that will be immediately transferrable to a wide array of
research fields. I have purposely written the book in a conversational style, foregoing a lot of strict mathematical rigor in favor of
intuitive logical explanations. Of course, mathematics does feature
throughout the book, but this is not a mathematics text. Indeed,
no knowledge of mathematics beyond high-school algebra and a
basic statistics course is assumed. Topics are covered in such a way
as to clearly communicate the fundamental principles of Bayesian
inference without getting lost in the nuances that can derail any
discussion of statistics. And most importantly, you will be provided
with many detailed examples of how to write up the results of your
Bayesian analyses.

Throughout the book, we will use the free software package
JASP, which can be downloaded from https://jasp-stats.org. JASP
was purpose-built for doing Bayesian inference, and its interface
will be familiar to people who already know some of the other popular statistics software packages. This is not a guidebook on using
JASP, though. Instead, JASP is the means to an end – our goal is to
figure out what Bayesian statistics is all about.

The book is designed to be read from start to finish at least once,
then you can go back and pick up topics again. Even if you already
know some of the early material, it is important to (at least quickly)
read through it so that you can see what perspective I am taking
throughout the book. If you work the examples along with me, you
will gain a lot of intuition about Bayesian inference.

One of my favorite self-descriptors is one I picked up a long time
ago – the source has been long forgotten: "converts knowledge thus

gained into student-friendly fare". My "why" in life has always been to help people do hard things. I don't even remember exactly when or why, but I started learning about Bayesian inference shortly after finishing my PhD. My teachers were people who wrote those early papers I was reading, including Mike Masson, Jeff Rouder, and E.J. Wagenmakers. Jeff and E.J., in particular, have always been so generous to me with their time and support. If you end up getting something out of this book, it is simply because they picked me up so that I could see over the fence. In essence, I have stood on *their* shoulders, converting the knowledge I gained from them into just the right form for *you*.

There are many people to thank who have contributed directly to this book in one way or another. I am grateful to Fred Oswald, Scott and Charla Bailey, Krzysztof Cipora, Mark LaCour, Curt Carlson, Bryanna Scheuler, Amy Bohmann, and Blake Wood, each of whom reviewed an early version of the book and provided many helpful comments and suggestions. I am grateful to the students at Tarleton State University who enrolled in my Bayesian statistics course in summer 2024 and took a chance to learn some things while I was writing the early manuscript: Hannah Abelein, Nicole Bloch, Ty Cosper, James George, Alexander Hoxie, Lauren Mayo, Samantha Mcgovern, Autumn Patterson, Michaela Plowman, Al Plucker, Beau Purdom, Addison Singleton, Kaileigh Smith, Ryon Springer, and Lindsey Willingham. I am thankful for the many conversations with my undergraduate and graduate research students who have followed me down the Bayesian path early in their career, including Kristen Bowman, Keelyn Brennan, Bryanna Scheuler, Landry Smith, and Bella Zapata. Also, I have a lot of great colleagues (including Amy Bohmann, Andrew Conway, Jo-Anne LeFevre, and Victoria Simms) who keep enabling me to teach and share my love of Bayesian inference by inviting me to give workshops.

Finally, I could not have even dreamed of writing a book like this without the love and support of my wonderful family.

I hope you enjoy this book and can feel the joy I had while writing it.

REVIEW OF BASIC CONCEPTS

Welcome to *Bayesian Statistics: The Basics*! I am so glad that you are joining me in taking your first steps along a rewarding path to learning the basics of Bayesian statistics. My goal in writing this book is to introduce you to the basic concepts of Bayesian statistics, particularly as it is applied to performing statistical inference in the most common research designs found across a wide array of empirical science disciplines, including psychology, neuroscience, animal science, biology, and others. Along the way, you will learn *what* we mean by a "Bayesian" approach to inference, as well as *why* we might want to use such an approach. Importantly, you'll learn *how* to perform Bayesian inference using the free software package JASP (https://jasp-stats.org). More about that soon.

The goal of this first chapter is to serve as a brief review of the basic statistical concepts that most people know before becoming motivated to learn Bayesian statistics. We will call this approach the *classical* (or *frequentist*) approach, though it is referred to in other places as the Neyman-Pearson approach and/or null hypothesis significance testing. This will be a very condensed review, of course, and anyone desiring more details is welcome to consult another book on statistics. In particular, my previous book in this series, *Psychological Statistics: The Basics*, will give you a quick and accessible primer on the basics of statistical inference. As such, this chapter will not really get into Bayesian statistics yet, but rather it will set the stage for *why* we might want to exert some effort in reading the remainder of this book.

Before moving into our review of classical statistical inference, I want to say a quick word about my writing style. I enjoy writing

DOI: 10.4324/9781003470199-1

books like these, as it gives me a license to write as I would speak in a classroom. This usually results – I am told, anyway – in a relaxed, conversational style. My hope is that when you read my words, you will get the experience of hearing me talk *with* you about Bayesian statistics. Friendly and approachable is always my goal when writing, so I hope that you feel this during your time spent with this book.

With that said, let's get started.

INFERENCE FOR CORRELATIONS: MEASURING PERSONALITY

Throughout this book, I will always begin any statistical discussion with a working example. For this chapter, our working example will focus on measuring the various characteristics of *personality*. Many times, the examples will be from my own discipline (the behavioral sciences), but not always. But please be aware that the concepts we will talk about transcend any specific discipline – they apply to so many more fields of scientific inquiry than I can begin to mention. A good mental exercise for you as you read this book is to think about examples that are closer to your own field of interest and imagine how you could apply the concepts discussed to those examples.

Here is the example we will work with in this chapter. The *NEO Personality Inventory-Revised* (abbreviated as NEO PI-R; *Costa & McCrae, 1992*) is a 240-item questionnaire that measures five dimensions of personality: Agreeableness, Conscientiousness, Neuroticism, Extraversion, and Openness to Experience. Usually referred to as the "Big Five", these personality dimensions can be easily remembered by use of an easily remembered acronym, for example, OCEAN or CANOE. The items in the NEO PI-R are a set of statements for which the respondent can indicate agreement using a five-point scale. Each item is associated with exactly one of the five dimensions. For example, to get a score for the dimension of Extraversion, the scores on the items that are uniquely associated with Extraversion are averaged to provide a composite score. This is done for each of the other four dimensions, so that the individual being measured by the NEO PI-R gets a set of five composite scores – one score for each of the five personality dimensions.

Once we have an individual's set of NEO PI-R scores, how should we interpret them? Here are some basic guidelines.

- Individuals scoring high in **Agreeableness** tend to value social harmony and getting along with others, whereas individuals scoring low in Agreeableness are generally prone to value self-interest over getting along with others.
- Those scoring high in **Conscientiousness** tend to have self-discipline and strive for achievement, whereas people scoring low in Conscientiousness tend to lack impulse control and are generally spontaneous.
- Individuals scoring high in **Neuroticism** tend to have strong negative emotions, whereas those scoring low in Neuroticism tend to be less emotionally reactive.
- People scoring high in **Extraversion** tend to create energy from external means and enjoy interacting with people. Those who are low in Extraversion tend to be quiet, deliberate, and less interested in social situations.
- Those scoring high in **Openness to Experience** tend to be intellectually curious and willing to try new things. Those scoring low in Openness to Experience are often pragmatic and less creative.

One question that could be asked about NEO PI-R scores is the following: *To what extent are the five dimensions of personality associated with each other?* That is exactly the question we will pursue in this chapter. We will use this work to review the basic concepts of classical statistical inference, which will in turn expose some fundamental limitations to the classical method. These limitations will motivate our discussion of Bayesian methods that begins in Chapter 2.

But first, we need two things: (1) some data; and (2) a way to work with it. Fortunately, both needs are satisfied by a free software package called *JASP*, which we will now discuss.

THE JASP SOFTWARE PACKAGE

JASP is a statistical software package that is freely downloadable from https://jasp-stats.org. It was developed by a team of researchers from the University of Amsterdam, and it is quickly becoming a popular tool for both teaching and research. It has several advantages over other popular statistical software packages, like SPSS and R. First, it is *free* and *open-source* – not only does it require no cost to use (other than having a computer to run it on, of course), but its entire source code is open for others to view and modify. Second, it is very *friendly* to users (especially new ones) – rather than requiring knowledge of programming, it has a graphical interface that was designed to be immediately usable and intuitive. Finally, it is *flexible* – users can perform standard analysis procedures in both classical as well as Bayesian versions.

Free, friendly, and flexible – this makes JASP the ideal tool for us to learn the basics of Bayesian statistics!

Downloading JASP is easy. On the JASP homepage (https://jasp-stats.org), there are multiple places to click that say something like "Download JASP". Any of these links will take you to the download page (https://jasp-stats.org/download), which will present you with several options for download packages. At the time I am writing this chapter (mid-2024), there are three operating systems that are supported: Windows, macOS, and Linux. Instructions for each are documented very well on the JASP website, as well as system requirements. Note that tablet computers are not currently supported; you'll need a traditional computer (desktop or laptop) to run JASP. Finally, I recommend that you install the most up-to-date version of JASP that your machine will support.

Once you download and install JASP on your computer, go ahead and open it. You'll be greeted with a screen that looks something like Figure 1.1.

The first thing we'll do is load a dataset into JASP so that we can begin our work. The data we will use comes from *Dolan, Oort, Stoel, and Wicherts (2009)*, who administered a Dutch version of the NEO PI-R to 500 first-year psychology students at the University of Amsterdam. This dataset is included in the JASP Data Library; that is, it is already built into JASP. While any data file in text form (e.g., a standard comma-separated or tab-delimited file) can be loaded into JASP, the most convenient thing for us to do at first is to use these datasets that are included in the JASP Data Library. To access this library and load our dataset, you'll first want to click the Main Menu button, which appears as three horizontal lines at the top left of the upper menu bar. When you click that button, opening the main menu, select "Open", then "Data Library". From here, you'll select the folder named "4. Regression". Figure 1.2 shows the result.

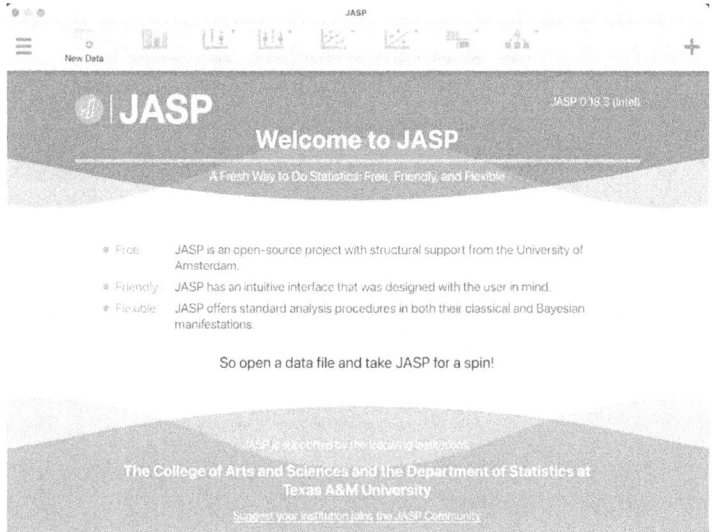

Figure 1.1 The "splash" screen that appears when JASP is opened.

Here, you will see a list of datasets – one of these is called "Big Five Personality Traits", and that is the one we want. Note that there are two icons that look like spreadsheets. One of the icons will have a little JASP logo on it, whereas the other one will not. The difference between these two icons is that the one with the JASP icon will contain a bunch of pre-defined JASP analyses, thus showing the user examples of the kinds of analyses that can be run on that dataset. The one without the JASP logo will contain just the dataset alone. This is the one we want, so go ahead and click on the icon that does *not* have the JASP logo. This will open the NEO PI-R data from *Dolan et al. (2009)*, and it should look like Figure 1.3. As one might expect from our previous discussion, there are five columns, each containing one of the five dimensions of personality. If you scroll down through the dataset, you'll see that there are 500 rows – one for each of the 500 undergraduate psychology students to which *Dolan et al. (2009)* administered the NEO PI-R. The numbers you see are the composite scores for each student on each of the five dimensions, scaled to range from 1 (lowest) to 5 (highest).

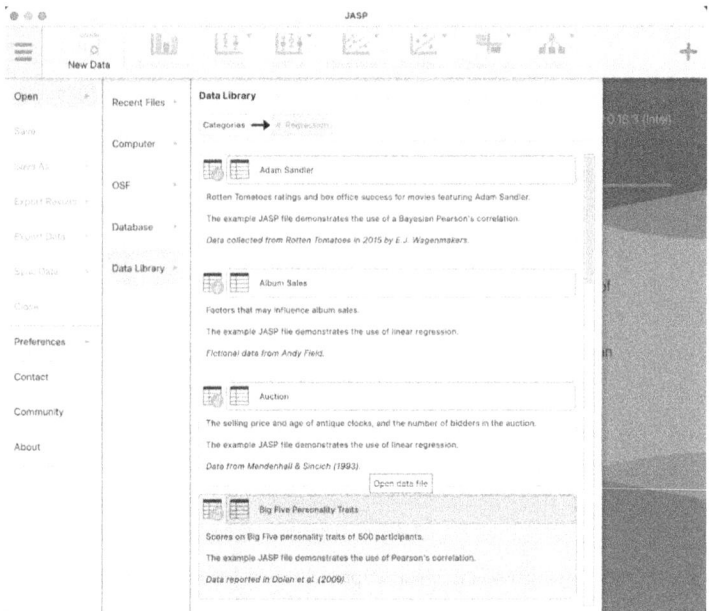

Figure 1.2 A screenshot of the JASP Data Library displaying the Big Five Personality Traits dataset.

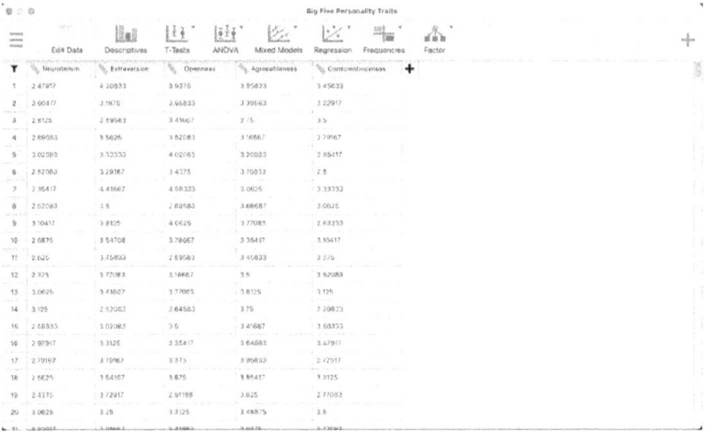

Figure 1.3 The "data screen" for the Big Five Personality Traits dataset in JASP.

Note that this data structure is the most typical type that you will use when performing your statistical inference in JASP. This data structure is often called "wide" format, where each row represents an individual unit of study (e.g., a person) and each column represents the various measurements that have been taken for that individual.

Let's now ask a specific question about these data: *To what degree is Neuroticism associated with Agreeableness?* As a first step, we can use JASP to run some descriptive analyses. To do this, click on the "Descriptives" button in the top menu bar. This will reveal the *analysis screen* – you'll move the two variables of interest (Neuroticism[1] and Agreeableness) into the "Variables" box by highlighting each variable name and then clicking the right-facing arrow to move it into the box. Additionally, we'll want to plot a "Correlation plot" – this is easily done by opening the "Basic Plots" menu and selecting "Correlation plots". The setup should look like Figure 1.4.

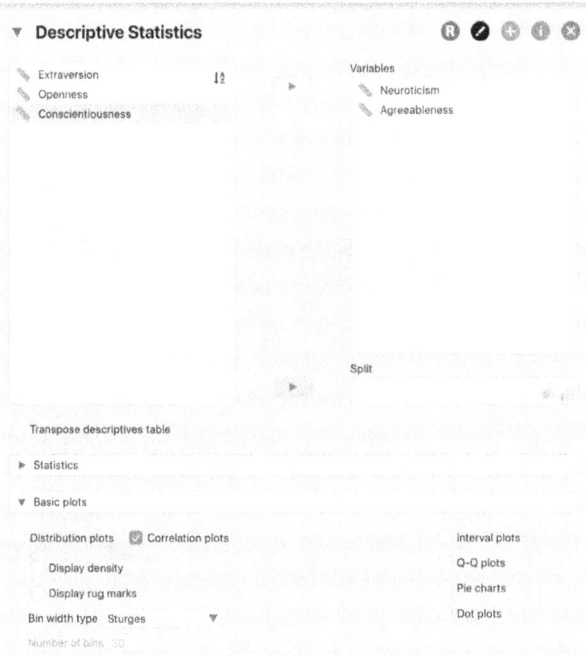

Figure 1.4 The descriptive statistics analysis screen in JASP.

Once you select these options, you may be surprised that the right side of your JASP screen is already busy producing the results of these analyses. In fact, that is one of the pleasing features of JASP that most new users comment on – there is no need to click a "go" button. All analyses proceed quickly and automatically appear in the *results screen* on the right side of the display.

In this example, you should see two main outputs. First, there is a table of descriptive statistics, containing things like the mean and standard deviation (note – you can select more descriptives to show by opening the "Statistics" menu and selecting any that you would like to see). Second, there is a correlation plot showing both a histogram of each selected variable (Neuroticism and Agreeableness) as well as a scatterplot showing their association (see Figure 1.5). Here, we can immediately see that (1) both variables are approximately normally distributed; and (2) there

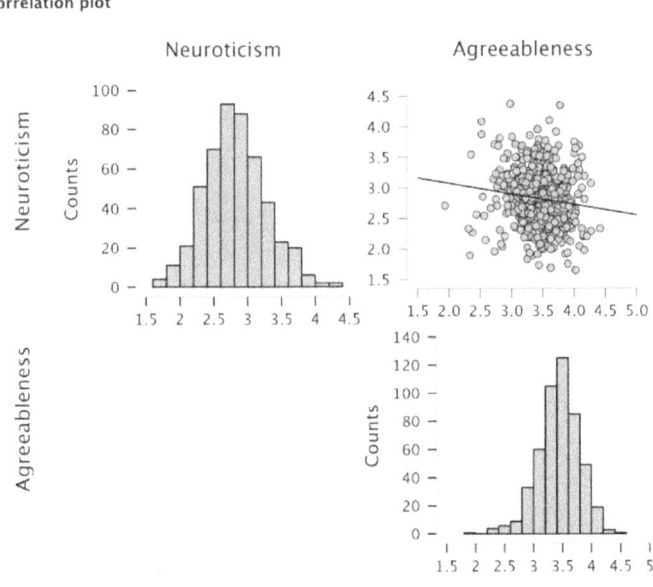

Figure 1.5 A JASP correlation plot that displays the relationship between Neuroticism and Agreeableness.

appears to be a slight negative relationship between `Neuroticism` and `Agreeableness`, implying that people scoring higher in `Agreeableness` (displayed on the horizontal axis) tend to score a bit lower in `Neuroticism` (displayed on the vertical axis).

Given these observations, our next step is to decide whether there is a *statistically meaningful relationship* between the variables `Neuroticism` and `Agreeableness`. I think this is a good time to have a quick review of how (classical) statistical inference works.

A QUICK REVIEW OF STATISTICAL INFERENCE

To decide whether these two personality dimensions are meaningfully related beyond the sample we are observing, we need to know what our sample of 500 students can say about the general population. One way to do this is to define two competing *models* about the general population. One of these models will propose that the two dimensions *are* related, whereas the other will propose that they are *not* related. Then, we will consider how well these models fit the *observed* data. I'll explain exactly what this means in a moment, but if you'll forgive me a bit of foreshadowing, this notion of assessing the fit of these competing models to observed data is one of the main ways in which Bayesian methods will differ from the classical methods you may already know.

To guide our discussion, consider Figure 1.6, which describes a framework for thinking about statistical inference that I proposed in an earlier book (*Faulkenberry, 2022*). This figure is an example of the *Kanizsa (1976)* triangle illusion, and I think it is a good framework for understanding how statistical inference works. There are three components that serve as the three anchors of the figure: (1) describing the data we observe, (2) defining some statistical models that represent competing views of reality, and (3) comparing the models and estimating the model parameters. Importantly, if you look at Figure 1.6 long enough, you may perceive a triangle that pops out of the middle between the three wedge-shaped anchors. It is important to note that the triangle is not physically there – rather, it is an inferred object of our perception. Further, the triangle only appears if the three wedge-shaped anchors are perfectly aligned. This illusory triangle serves as a nice metaphor for our role as scientists, as we seek out knowledge about things that are

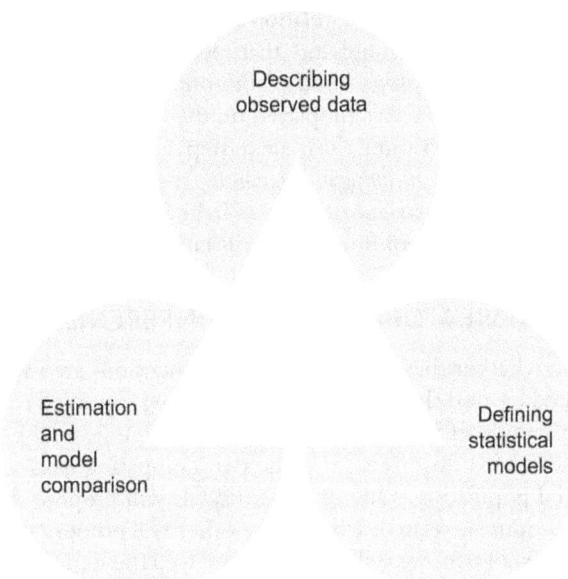

Describing
observed data

Estimation
and
model
comparison

Defining
statistical
models

Figure 1.6 A framework for statistical inference based on the Kanizsa (1976) triangle illusion.

usually impossible to see directly. From the three concrete anchors of describing our data, defining models, and comparing the models, we are able to magically see the components of reality that were previously invisible – in this concrete case, the (previously hidden) relationship between neuroticism and agreeableness.

The first anchor of the inference framework is describing data. Let's consider how we would do this. In this case, where the observed data is an *association* between variables, we can use the *Pearson correlation coefficient*. Mathematically, the correlation ρ (the Greek letter "rho") between two continuous variables X and Y is given by

$$\rho = \frac{\sigma_{XY}}{\sigma_X \cdot \sigma_Y},$$

where σ_{XY} is the *covariance* of X and Y, and σ_X and σ_Y are the standard deviations of X and Y, respectively. Covariance simply measures how two variables "co-vary". Specifically, covariance is computed by taking the differences between each data point and its respective mean, multiplying these differences for the two variables of interest, and then averaging the result. If the values tend to increase or decrease together, the covariance will be positive; if one increases while the other decreases, it will be negative. Further, dividing by the standard deviations does the remarkable job of standardizing the covariance so that the result is always between -1 and $+1$. As the correlation ρ approaches ± 1, the degree of association increases; that is, a correlation of $\rho = 1$ or $\rho = -1$ represents a perfect positive (or negative, respectively) correlation. On the other hand, a correlation of $\rho = 0$ would indicate that there is no relationship between X and Y.

Once we have computed the correlation between two variables, how do we then decide whether those two variables are meaningfully[2] associated? The classical method (also known as *null hypothesis statistical testing*) proceeds like this. First, we build two models (also known as *hypotheses*) that represent two different, competing versions of reality. One model, which we call the *null hypothesis*, states that there is no association between the two variables. That is, this null hypothesis proposes that the correlation coefficient ρ is equal to 0. Mathematically, we can write this model as:

$$\mathcal{H}_0 : \rho = 0.$$

The other model, which we call the *alternative hypothesis*, can take several forms, but at its simplest it simply states the opposite of the null hypothesis; that is, that there is some nonzero (i.e., positive *or* negative) association between the two variables. In this case, we would write the alternative hypothesis as

$$\mathcal{H}_1 : \rho \neq 0.$$

Given these two competing models – one where there is no association between the two variables, and another where there is an association – our next step is to decide which model best "fits". That is, we want to know which model does a better job of predicting the data we actually observed. The classical method for this model comparison is summarized in Figure 1.7, but here's how it

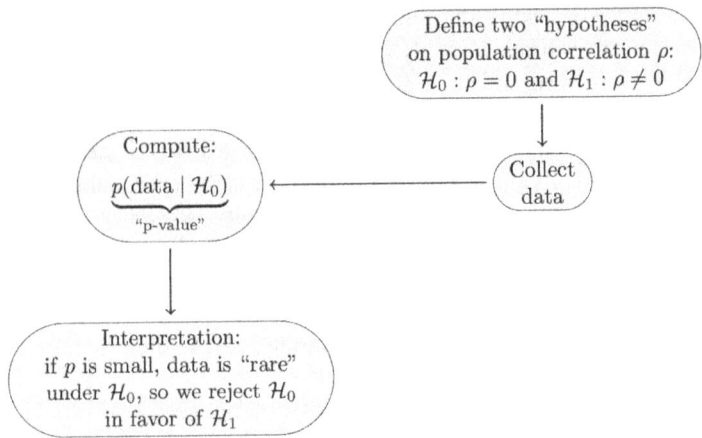

Figure 1.7 A flowchart of the logic involved in classical statistical inference.

works in a nutshell. First, we assume (just for the sake of argument) that the null hypothesis is true; that is, we assume $\rho = 0$. Then, we calculate the probability of observing our (actually observed) data under this assumption – that is, we compute the conditional probability $\rho\,(\text{data} \mid \mathcal{H}_0)$. If this probability (called the *p-value*) is small[3] (say, less than 5%), this implies that our observed data is actually rare under the null hypothesis. Simply put, we should not have observed that data by random chance alone. But we did . . . so we can conclude that the underlying assumption we made (namely, that the null hypothesis is true) must have been wrong. So we *reject* the null hypothesis; and given that we only have two models under consideration ($\mathcal{H}_0$ and $\mathcal{H}_1$), this leaves us with support for the alternative hypothesis $\mathcal{H}_1$.

Before moving on with our NEO PI-R example and performing a hypothesis test in JASP, I would like to take a moment to describe how the *p*-value is actually computed. Most readers will have likely seen a *p*-value as a standard output in some statistical software but may not have actually thought much about where that number comes from. As mentioned earlier, the *p*-value is (roughly speaking) the probability of observing the data in hand under the assumption that the null hypothesis $\mathcal{H}_0$ is true (that is, $\rho = 0$). This is *conceptually* right, but it is not technically 100% correct. As one of the arguments I will present later in this chapter against using p-values depends on it, we need to get this definition fully correct. So let's do that.

First, we need to know what it means to say, "compute the probability of observed data". What we really mean here is that we first convert the observed data into a number (called a *test statistic*) that represents the data in a way that is useful for our research and measurement context. In this case, the typical test statistic for comparing models about the association between two variables is the correlation coefficient for our sample[4] (e.g., $r = 0.073$), which is then transformed to a test statistic called a t-score (in this case, it happens to be that $t = 1.63$). We now want to know how likely this specific value of the test statistic is, compared to the distribution of values of the test statistic that we would expect if the population correlation $\rho = 0$. For this specific sampling situation, the distribution of these test statistics (sometimes called the *sampling distribution*) turns out to be a t-distribution. Knowing this, we can use the mathematics of the t-distribution (i.e., some calculus, or a table, or software, etc.) to compute the probability of observing test statistic values *more extreme* than $t = 1.63$. Because the alternative hypothesis $\mathcal{H}_1$ is nondirectional (i.e., $\mathcal{H}_1 : \rho \neq 0$), "more extreme" could be in either direction, implying that we need to know the probability of obtaining test statistic values *greater* than 1.63 or *less* than -1.63. Graphically, these probabilities can be viewed as "tails" of the t-distribution in Figure 1.8. Each tail is a little more than 5% of the distribution, so the sum of these two tails is just over 10% – giving $p = 0.103$.

With all these details discussed, we are now (finally) ready to see how to perform the correlation test in JASP. It is refreshingly easy. To access the correlation test, you'll want to click on the "Regression" button in the upper tool bar. This will drop down a few choices, among which you will choose "Correlation" from the Classical list. Don't worry – we'll do the Bayesian version in due time. This will produce a new analysis screen, though somewhat similar in feel to the one we got when we did descriptives earlier. Go ahead and move Neuroticism and Agreeableness into the "Variables" box. You'll immediately see some results populate on the right side of the screen, but let's choose a few more options before diving into the results. Additionally, go ahead and select

Distribution of test statistics under null

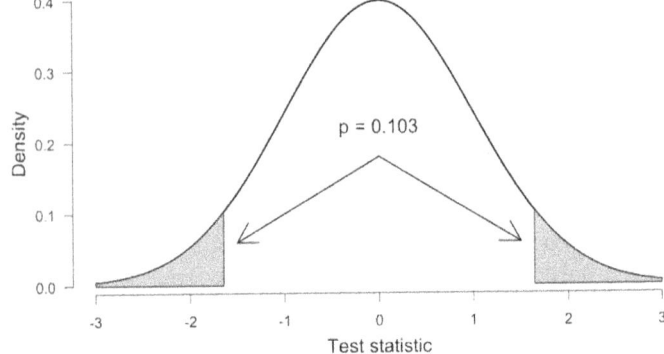

Figure 1.8 The distribution of values of the test statistic that would be expected if the null hypothesis was true. The shaded region displays those at least as extreme as the *observed* test statistic.

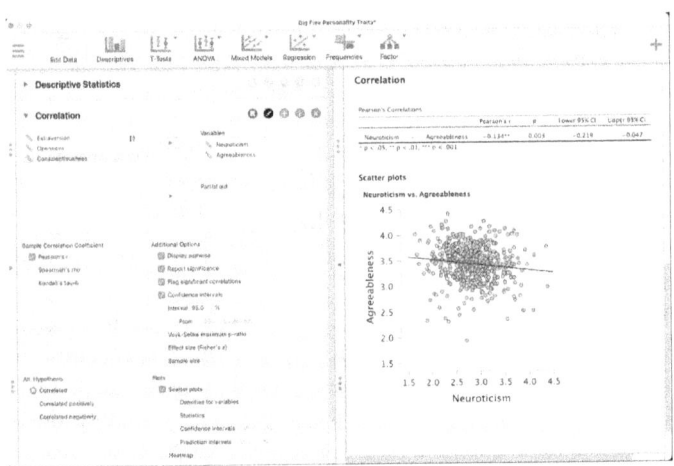

Figure 1.9 A JASP screenshot displaying how to perform a (traditional) correlation test to assess the relationship between Neuroticism and Agreeableness.

the following options: "Display pairwise," "Flag significant correlations", "Confidence intervals", and "Scatter plots". Your result should be similar to Figure 1.9.

In this results screen, we can see that there is a small, negative, statistically significant correlation between Neuroticism and Agreeableness, $r = -0.134$, $p = 0.003$. Moreover, the 95% confidence interval for the population correlation ρ is given by $[-0.219, -0.047]$. For review, let's quickly discuss what each of these things mean. First, $p = 0.003$ means that if we assume the null hypothesis $\mathcal{H}_0$ is true, then the probability of observing a correlation at least as extreme as $r = -0.134$ is pretty small – namely, $p = 0.003$. Thus, our data is very *rare* under $\mathcal{H}_0$, leading us to reject $\mathcal{H}_0$ as a potential model. This leaves us with (albeit indirect) support for $\mathcal{H}_1$, which states that there is some nonzero correlation between Neuroticism and Agreeableness.

Second, let us review what is meant by a 95% confidence interval. We often interpret a 95% confidence interval in the following way: "We are 95% confident that the population correlation ρ lies between -0.219 and -0.047". While it is perfectly fine to say this, we should be careful to indicate what this is *not* saying. This is not saying that there is a 95% chance of ρ falling between those two values. Indeed, the confidence interval does not give us any way to estimate ρ via a direct probability statement. Rather, a confidence interval is the result of a process based on the observed sample that, 95% of the time, constructs an interval which contains ρ. This is not quite the same thing as the previous statement. Indeed, confidence intervals are extremely counterintuitive and difficult to interpret correctly, even for seasoned researchers (*Hoekstra, Morey, Rouder, & Wagenmakers, 2014*). This limitation will be overcome in the next chapter when we introduce Bayesian methods.

As you might be starting to discern, there are some issues that arise when using these classical methods. Indeed, that is the direction I am intending to take this discussion, but we need to see one more example before we can really start to outline a convincing case against the classical method for inference.

For our next example, let us ask whether there is a meaningful association between Extraversion and Conscientiousness. We now have sufficient background and experience to easily test this in JASP; just move the variables from the previous example out of the "Variables" box and replace them with the new ones (Extraversion and Conscientiousness). The updated results screen is displayed in Figure 1.10.

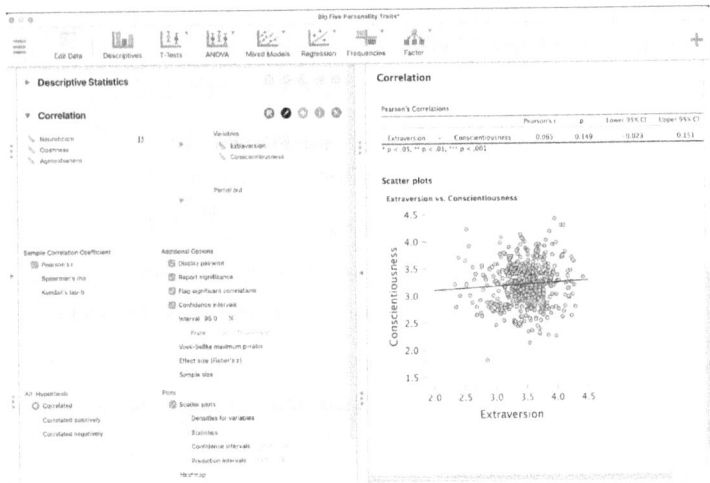

Figure 1.10 A JASP screenshot displaying how to perform a (traditional) correlation test to assess the relationship between Extraversion and Conscientiousness.

In this example, we can see that the correlation between Extraversion and Conscientiousness is not statistically significant, $r = 0.065$, $p = 0.149$, and the confidence interval $[-0.023, 0.151]$ contains $\rho = 0$ as a possibility. Again, let's walk through the logic of the hypothesis test. If we assume $\mathcal{H}_0$ is true, then the probability of observing a correlation at least as extreme as $r = 0.065$ is $p = 0.149$. Since this p-value is greater than 0.05, we consider these data to be somewhat *plausible*. In this case, the observed correlation is somewhat reasonably predicted by the null hypothesis, so we cannot reject $\mathcal{H}_0$ as a potential model.

It is perhaps interesting to note the following. In neither of these examples do we explicitly assess the fit of the data with the alternative hypothesis $\mathcal{H}_1$. In the first example, we choose $\mathcal{H}_1$ simply because $\mathcal{H}_0$ did not do a good job of predicting the observed correlation. But in the second example, we cannot reject $\mathcal{H}_0$ because it does a decent job of predicting the observed correlation. What about $\mathcal{H}_1$? Unfortunately, the classical method of inference does not give us a way to measure how well $\mathcal{H}_1$ predicts the observed correlation. Thus, we are left wondering what to do, as we cannot make a clear decision. This conundrum reveals another fundamental limitation of classical null hypothesis testing – a nonsignificant p-value (i.e., $p > 0.05$) does not lend support for $\mathcal{H}_0$.

SOME ARGUMENTS AGAINST THE USE OF p-VALUES

So far in this introductory chapter, we have gone through a review of classical inference. Specifically, we have reviewed using null hypothesis testing with p-values as the primary model comparison index. However, the discussion immediately preceding this section has hopefully cast some doubt in your mind about whether this approach is the best one we can use. Of course, given that this is a book on Bayesian statistics, that is exactly the fork in the road that I am trying to lead you to take. In this final section, I will close with a set of three arguments against the traditional use of p-values as a tool for scientific inference. These arguments were originally put forth by *Wagenmakers (2007)*, and I believe they are still some of the best arguments for considering an alternative method of inference. Please note – these arguments are not against the p-value itself. For sure, p-values are perfectly well-defined mathematical objects that tell us *something* about our observed data. The argument is simply that they tell us something that isn't really as useful as our past statistical training and scientific practice may have led us to believe.

ARGUMENT 1 – p-VALUES DEPEND ON DATA THAT WERE NEVER OBSERVED

Early in the chapter, I gave a conceptual definition of the p-value that went something like this: "The p-value is the probability of observing some data, given that the null hypothesis is true". Of course, a bit later I admitted that this is not quite technically correct, as the formal definition of the p-value is really the probability of obtaining a given test statistic (derived from the observed data), *or a more extreme value of the test statistic*, given that the null hypothesis is true. This necessary modification to the definition of the p-value leads directly to the first argument against it. As we saw in our earlier example, the p-value for an observed correlation of $r = 0.073$ is represented by the tail regions of a t-distribution. Most of the mass in this pair of tail regions consists of *hypothetical* data that were never actually observed. In fact, in Figure 1.11, we can see that

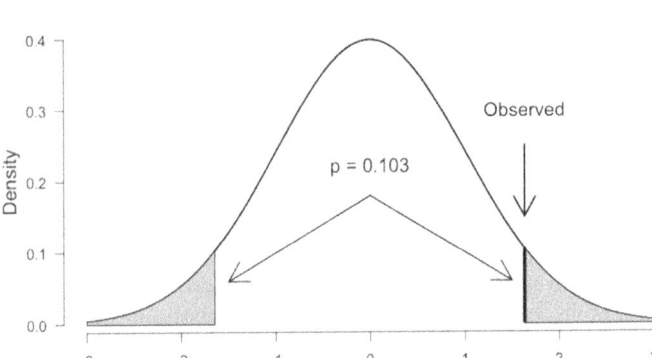

Figure 1.11 The (actually) observed data is just a slice of the shaded region used to compute a *p*-value.

the actually observed data ($r = 0.073$, transformed to $t = 1.63$) only forms one (infinitely thin) slice of the area that goes into computing the *p*-value.

Why is this a problem? In his original paper, *Wagenmakers (2007)* gives several reasons why depending on data that were never observed is less than optimal. One particularly compelling reason is that such actions violate two fundamental philosophical positions in statistics known as the *conditionality principle* and the *likelihood principle (Birnbaum, 1962)*. While the mathematical formulations of these principles are beyond the scope of this book, Birnbaum himself gives informal descriptions of these two principles. Together, these two principles assert that experiments and outcomes not actually observed are not relevant to informative inference. A direct implication of this argument, then, is that we should seek model comparison metrics that depend only on the data we actually observe.

ARGUMENT 2 – *p*-VALUES DEPEND ON A RESEARCHER'S (POSSIBLY UNKNOWN) SUBJECTIVE INTENTIONS

When we observe a particular result (say, $r = 0.073$) with given sample size, it seems straightforward to go through the classical ritual and compute the associated *p*-value as $p = 0.103$. *But Wagenmakers*

(2007) gives a simple example that may serve to cast some doubt on how simple the situation really is. The example goes like this – suppose you are given data from a survey of 12 items, each of which is a factual true/false question (so that the probability of getting any one question correct is exactly 0.50). Suppose further that upon scoring this survey, you see that the taker has gotten 9 of the 12 questions correct, a "passing" score of 75%. You are then asked whether it is possible that the test taker simply guessed. As a classically trained researcher, you decide to perform a hypothesis test, using a *binomial distribution* as the underlying statistical model. We'll talk more about the binomial distribution in the next chapter, but for now, it suffices to know that this model is perfect for situations where you have some number of items with only two outcomes. The critical parameter of interest is θ, which is the probability of getting any one item correct. If someone is just guessing, then this would imply $\theta = 0.50$. Thus, we set our null hypothesis as $\mathcal{H}_0 : \theta = 0.50$, and our alternative hypothesis as $\mathcal{H}_1 : \theta > 0.50$. We observed nine correct answers, so the *p*-value would be computed by first assuming the null (i.e., that $\theta = 0.50$) and then calculating the probability of obtaining nine correct answers or more – that is, 9, 10, 11, or 12 correct answers). Some high-school algebra (or looking ahead in Chapter 2 for the binomial formula) can be employed to compute this probability, which is $p = 0.0729$. Since this *p*-value is greater than the customary threshold of 0.05, we fail to reject the null hypothesis, from which we reply that we cannot rule out the possibility that the test taker was simply guessing.

However, the model we used (the binomial model) assumes that the researcher intended to give a total of 12 questions, from which the test taker incorrectly answered 3. Could it be possible that the researcher instead intended to simply give questions *until* the test taker missed three of them? If this were the case, the binomial model would no longer be appropriate. Instead we would need to use something called the *negative binomial model*, as it is based on stopping at the desired number of successes. Again, details are not important here, but the *p*-value one would obtain if using the negative binomial model would be quite different – in this case, $p = 0.033$. Following conventional rules about *p*-values would lead us to reject the null in this case, thus ruling out the possibility that the test taker was guessing.

The point of this discussion is that given a set of observed data, we do not necessarily know what the subjective intentions of the researcher were when the data were collected. The previous example illustrates that two different sampling plans (i.e., two different "subjective intentions") result in very different p-values. Of course, this not a fault of the p-value itself, but rather an important warning about using p-values – namely, that the underlying statistical models should be known and specified. This is rarely the case in most traditional applications of classical null hypothesis testing.

ARGUMENT 3 – p-VALUES DO NOT QUANTIFY STATISTICAL EVIDENCE

In our most recent JASP example, we found that the correlation between extraversion and conscientiousness was small, giving us a p-value which indicated that such data could in fact be *plausible* under the null hypothesis $\mathcal{H}_0 : \rho = 0$. Thus, we failed to reject the null hypothesis. Commonly, such a result is interpreted as evidence *against* the alternative hypothesis, but it is important to keep in mind that failing to reject the null does not necessarily imply evidence *against* the alternative hypothesis. Indeed, the ability of $\mathcal{H}_1$ to predict the observed data has not actually been measured in any way. Beyond this one case, I believe this is a critical flaw of the traditional null hypothesis testing procedure. Given two competing models $\mathcal{H}_0$ and $\mathcal{H}_1$, the only one that is "tested" against observed data is $\mathcal{H}_0$. Whenever we decide to reject $\mathcal{H}_0$ because of poor predictive ability (i.e., a small p-value indicating that the observed data we have actually should be quite rare under $\mathcal{H}_0$), we then "rule" in favor of $\mathcal{H}_1$. This favorable choice of $\mathcal{H}_1$ is not due to any formal assessment of how well $\mathcal{H}_1$ predicts the data we have, but rather because it is the only alternative. It would be similar to having an election where the winner was chosen by picking the person who had the least number of "against" votes.

Beyond this, it is also true that p-values do not coherently quantify statistical evidence. *Wagenmakers (2007)* gives the following argument. Consider two experiments to test some empirical phenomenon. In Experiment 1, researchers tested 11 subjects and found $p = 0.032$, whereas in Experiment 2, researchers tested 98 participants and also found $p = 0.032$. Which experiment provides more "evidence" for the phenomenon in question? Or is the amount of evidence the same between the two experiments?

This is, in fact, quite a complicated question, as its answer might very well depend on what we mean by "evidence"? Certainly, this appears to be the case, as multiple answers have been offered in the literature. For example, *Rosenthal and Gaito (1963)* found that psychologists were more likely to reject the null hypothesis as sample size increased. In this case, that would imply that these psychologists would intuitively pick Experiment 2 as the one that provides more evidence. On the other hand, others (e.g., *Lindley & Scott, 1984*) have argued the reverse – for a fixed value of p, the experiment with the *smaller* sample provides more evidence. One reason is that the p-value depends on both the sample size and the effect size of the observed data. If p is fixed to be constant, then increasing the sample size in Experiment 2 must necessarily decrease the observed effect. Thus, the observed effect is smaller in Experiment 2 than in Experiment 1, which gives the edge to Experiment 1 as the one with greater evidence for the phenomenon in question. Regardless of which argument you believe, the key takeaway is that the observed p-value does not coherently measure the statistical evidence provided by the observed data.

CHAPTER SUMMARY

Before moving on to Chapter 2, let's recap what we've learned in this chapter:

1. We reviewed the basics of classical statistical inference, which involves null hypothesis testing.
2. We introduced the free software package JASP, which is an easy-to-use tool for doing both classical and Bayesian inference.
3. We walked through a couple of examples of performing correlation tests.
4. We went into great detail about what p-values are, and critically, what they are *not*.
5. We ended by considering three arguments against the use of p-values as a tool for statistical inference.

EXERCISES

1. Using the Big Five dataset from this chapter, consider the degree of association between Extraversion and Agreeableness.
    a. Write formal definitions of the two models to be compared in a hypothesis test ($\mathcal{H}_0$ and $\mathcal{H}_1$).
    b. Calculate the Pearson correlation coefficient and report the p-value.
    c. What does this p-value tell you about your two models? Which model do we prefer after observing the data?
    d. Report a 95% confidence interval for ρ.
    e. What can you conclude from these analyses?
    f. What happens to the p-value when you select "Correlated positively" from the "Alt. Hypothesis" menu? Try to explain why the p-value changes the way it does.
2. Open the "College Success" dataset in the JASP Data Library (you'll find it in the same collection as the Big Five data). Use the correlation test to answer the following question: *Which is the better predictor of college success (as measured by GPA): the math SAT score, or the verbal SAT score?* Be sure to report the models being compared, the test statistic(s) of the observed data (i.e., the Pearson correlation[s]), and the p-value(s). What do you conclude?

NOTES

1 Throughout the book, specific variable names will be indicated by using a `fixed width font`.
2 That is, that there is a correlation present in the population that goes beyond the sample itself.
3 Note that what is meant by "small" depends on the field. For many fields, 5% is the typical threshold, but some (including physics) use much smaller thresholds (e.g., *Lyons, 2013*).
4 As is customary, we will use the convention of representing population *parameters* with Greek letters (e.g., ρ) and parameter *estimates* with Latin letters (e.g., r).

2

THE LANGUAGE OF
BAYESIAN STATISTICS

In Chapter 1, we began our journey by reviewing the basics of classical statistical inference, with particular focus on the method of null
hypothesis testing. This method involves computing the p-value,
which is defined as the likelihood of obtaining your observed data
(or more "extreme") under the null hypothesis. Small values of p
(e.g., $p < 0.05$) lead us to *reject* the null hypothesis, which gives us
indirect evidence for the alternative hypothesis. We concluded the
chapter with a discussion of three limitations that arise when using
p-values for inference. First, p-values depend on data that were
never actually observed, which violates some fundamental philosophical principles in statistical inference. Second, p-values depend
on a researchers' possibly unknown subjective intentions, particularly related to how the observed data were actually sampled (i.e.,
was the sample "fully" obtained, or did sampling stop because a
particular value of p was obtained). Finally, p-values do not provide
a coherent way to quantify statistical evidence for either the null or
the alternative hypothesis.

In this chapter, we will introduce the basics of Bayesian inference and present it as an alternative to the classical method. At the
end, we will discover something to use for hypothesis testing that
does not suffer from the limitations described previously – the *Bayes
factor*. But perhaps more important is the journey we take along
the way. After reading this chapter, you will learn the *language* of
Bayesian statistics. Words like *prior, posterior, predictive*, and *marginal*
are used frequently in the context of Bayesian methods, and it is
important to gain a conceptual understanding of what these words
mean. As we do throughout this book, we will use JASP as our

DOI: 10.4324/9781003470199-2

primary learning tool. Usually, explanations of these concepts will be verbal in nature, supplemented by what we see in our JASP output. However, when appropriate, I will also present (and explain) mathematical formulas. To be clear, one way this book differs from the many other good books on Bayesian statistics is that we rely much less on having a deep mathematical background and focus instead on conceptual descriptions, letting software handle (most of) the mathematical bits. With this in mind, I'll try to keep the equations palatable and minimal.

WHAT IS BAYESIAN STATISTICS?

Perhaps a good place to start would be to go ahead and give a definition of Bayesian statistics. I will admit that this is a difficult task, but let's give it a shot.

At its core, *Bayesian statistics* is all about learning from data. Formally, it is a system of statistical inference that uses prior knowledge to *predict* (potential, pre-observation) data, and after data is actually observed, uses the error in the prediction to *update* the prior knowledge. This new *posterior* knowledge can then be used to make further predictions about new data, from which the cycle can be repeated again and again (see Figure 2.1). It is a fully mathematical approach based on probability theory (specifically, Bayes' theorem – more about this in a minute). Moreover, this Bayesian learning cycle mirrors exactly what we do in science. We use what

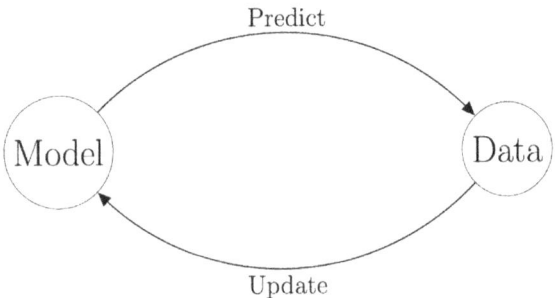

Figure 2.1 The Bayesian learning cycle – prior knowledge about a model is used to predict data, and observed data is to construct posterior knowledge and update the model.

we already know to make new predictions that we can test, and then we use the observed data to update what we know.

The mathematical foundation of Bayesian statistics comes from *Bayes' theorem*, sometimes called "Bayes' rule". The theorem itself is easy to write down. In the context of two "events" A and B, Bayes' theorem says

$$p(A \mid B) = p(A) \times \frac{p(B \mid A)}{p(B)}.$$

Without context, this statement gives us a simple mathematical fact about *conditional probabilities*. It says that the conditional probability of A given B (i.e., the probability of event A happening, provided we know that event B has occurred) is equal to the probability of A multiplied by the conditional probability of B given A, divided by the probability of B.

Incidentally, Bayes' theorem itself is very easy to prove. All we need to know is that the conditional probability $p(A \mid B)$ is calculated by taking the probability of the joint event $A \cap B$ and dividing by the probability of the single event B. We can use this definition to derive the following two equations:

$$p(A \mid B) = \frac{p(A \cap B)}{p(B)},$$

and

$$p(B \mid A) = \frac{p(B \cap A)}{p(A)}.$$

Since the joint event $A \cap B$ (i.e., A "and" B) can be equivalently written as $B \cap A$, we can solve the second equation for $p(B \cap A)$:

$$p(B \cap A) = p(A) \times p(B \mid A),$$

and substitute that into the first equation for $p(A \cap B)$. Doing this substitution, along with some simple statements in elementary set theory, gives us Bayes theorem as a result:

$$p(A \mid B) = \frac{p(A \cap B)}{p(B)}$$

$$= \frac{p(B \cap A)}{p(B)}$$

$$= \frac{p(A) \times p(B \mid A)}{p(B)}$$

$$= p(A) \times \frac{p(B \mid A)}{p(B)}.$$

One very common (and useful) application of Bayes' theorem is to assess the positive predictive value of a medical test. For example, suppose that during a doctor's visit, a quick test for influenza comes back positive. What is the probability that you *actually have the flu?* This question can be posed as a conditional probability: $p(\text{flu} \mid \text{positive})$. Since this question is posed as a conditional probability, we can immediately apply Bayes' theorem:

$$p(\text{flu} \mid \text{positive}) = p(\text{flu}) \times \frac{p(\text{positive} \mid \text{flu})}{p(\text{positive})}.$$

To work out the answer, we need to know what each of the components on the right side of this equation mean. Let's work through them one at a time:

1. $p(\text{flu})$: This unconditional probability represents the *prior probability* of a patient having the flu, before seeing any test result. In clinical applications, this prior probability is often called the "prevalence" of the disease, and it simply refers to the probability that a randomly chosen person from the population has the flu. This is equivalent to the proportion of people in the population who have the disease at a given moment of time. Let's suppose, for the purposes of our example, that we are in late summer, and the prevalence of influenza is reasonably low, say 10%. Then we would assign our prior probability as $p(\text{flu}) = 0.10$.

2. $p(\text{positive} \mid \text{flu})$: This conditional probability represents the *likelihood* of observing a positive test, given that the patient actually has the flu. In clinical applications, this likelihood is often called the "sensitivity" of the test. The value of this sensitivity depends on the test itself, but note that the United States Food and Drug Administration requires all of these typical "rapid" diagnostic tests to have a sensitivity of at least 80% before issuing approval for use. So let's assume that the test in our example exceeds this criterion; that is, $p(\text{positive} \mid \text{flu}) = 0.90$.

3. $p(\text{positive})$: This represents the *marginal* probability of observing a positive test. Key to computing this probability is noting that there are two paths that we could follow to observe a positive test. The first path is to observe a positive test from someone who actually has the flu. The second path is to observe a positive test from someone who *doesn't* have the flu (i.e., a false positive). Thus, to compute this probability, we need to compute the probability of each of these paths. For the first path, we compute

$$p(\text{positive} \mid \text{flu}) \times p(\text{flu}) = 0.90 \times 0.10 = 0.09.$$

For the second path, we need to know the probability of observing a false positive; let's assume for our example that this test has a false positive rate is 8%, so that $p(\text{positive} \mid \text{no flu}) = 0.08$. Combining this with the prior probability of *not* having the flu (i.e., $1 - 0.10 = 0.90$), we can compute the probability of observing a false positive:

$$p(\text{positive} \mid \text{no flu}) \times p(\text{no flu}) = 0.08 \times 0.90 = 0.072.$$

Thus, the marginal probability is given by the sum of these two paths:

$$p(\text{positive}) = 0.09 + 0.072 = 0.162 \, .$$

Now we are ready to answer our question. The probability of actually having the flu, given that we observed a positive test, is

$$p(\text{flu} \mid \text{positive}) = p(\text{flu}) \times \frac{p(\text{positive} \mid \text{flu})}{p(\text{positive})}$$

$$= 0.10 \times \frac{0.90}{0.162}$$

$$= 0.556.$$

Thus, even though we observed a positive test, the probability of actually having the flu is only 0.556. It is perhaps counterintuitive that a positive result on a test with 90% sensitivity can lead to such a low probability of actually having the flu. Of course, the result comes from combining this sensitivity with several other pieces of information, including the prior probability of having the flu and the false positive rate of the test. For example, what would happen if we were in winter and the prior probability of having the flu was greater than 0.10 − say, $p(\text{flu}) = 0.60$? In this case, running through the aforementioned calculations with this new value for $p(\text{flu})$ would produce a much larger posterior probability of $p(\text{flu} \mid \text{positive}) = 0.944$. Try some calculations like this for yourself and see if you can get some intuition for how it all works!

BAYES' THEOREM APPLIED TO DATA

When I was first learning about Bayesian statistics early in my career, I had a mistaken impression that it was all about the kinds of problems I just described. Honestly, that didn't seem too exciting to me, and it was not at all clear to me how we could apply this type of reasoning to hypothesis testing. What I was missing was the realization that Bayes' theorem has a very useful form that can more clearly be applied to the kinds of things that we are interested in when working with inferential statistics. This "data" form of Bayes' theorem looks like the following:

$$\pi(\theta \mid \text{data}) = \pi(\theta) \times \frac{p(\text{data} \mid \theta)}{p(\text{data})}.$$

Though the form and structure may appear similar to the version of Bayes' theorem I presented earlier, there are some differences in notation that need to be explained. First, the context we are working with in this version is that which we described in Chapter 1 − namely, of model parameters and data. Here, θ is the variable I am using to represent a model parameter (you'll see why pretty soon), but you could just as easily use as our parameter the correlation coefficient ρ from Chapter 1. Against this background, the components of Bayes' theorem are listed next.

1. $\pi(\theta \mid \text{data})$: This is called the *posterior distribution* for θ. *Posterior* means "after observing data". We use the notation π to indicate that it is a probability *density* (i.e., a positively-valued mathematical function whose area under the curve is equal to 1 over its domain). Whereas in the aforementioned example we were interested in computing the probability of a specific event (giving us an answer that is a single number between 0 and 1), under this context we are interested in the distribution of possible values for our model parameter θ.

2. $\pi(\theta)$: This is called the *prior distribution* for θ. Like the posterior distribution, we use the π notation. The difference is that this probability density function represents the distribution of values of θ *before* observing data. One of the big conceptual hurdles to learning Bayesian statistics is getting over the idea that priors are somehow just "made up" or completely subjective. Though it is possible to define any prior that one wishes, the approach I will describe throughout this book is to describe *objective* prior distributions that work well for the types of analyses we commonly perform. I'll be clearer about what it means to "work well" a bit later.

Rouder and Morey (2018) give a nice description of the role of $\pi(\theta \mid \text{data})$ and $\pi(\theta)$ in Bayes' theorem. They explain that these two distributions represent our beliefs about values of θ and, in particular, how these beliefs are updated in light of observing data. Because these beliefs are fluid and contain uncertainty, we must represent them with a distribution, and hence the π-notation. These probabilities are of a fundamentally different type than the other probabilities in Bayes' theorem, which I will describe next:

1. $p(\text{data} \mid \theta)$: This is called the *likelihood* function, and it represents the probability of observing the data given a specific value of θ. The likelihood function depends on the specific model that is being used to represent the data. We'll walk through an example of computing this likelihood momentarily.

2. $p(\text{data})$: This is called the *marginal* probability of the data. This quantity is often misunderstood, as its simplicity of form hides what it really is. The idea is that it is the weighted average of the likelihoods for the data over all possible values of the model parameter θ, where the weighting applied is the prior distribution for θ. For those mathematically inclined,[1] the definition is

$$p(\text{data}) = \int p(\text{data} \mid \theta)\pi(\theta)d\theta.$$

This equation says that the marginal probability of the data is obtained by summing (that's what the integral sign essentially means) the products of the likelihoods that you get by varying the values of the parameter θ multiplied by the prior density of each value of θ.

A SIMPLE EXAMPLE

At this point, I think it is imperative that we work through a simple example together so that we can get a feel for what is being described in Bayes' theorem. Please don't skip this – the reward for working through these details will be great and will pay large dividends for your conceptual understanding as we work through the later chapters.

Back in early 2024, I received an invitation to visit some colleagues at New Mexico State University and give a workshop on Bayesian statistics. As I always try to come up with examples that are unique to the location of my workshops, I attempted to find some data collection scenario that was unique to New Mexico. In the course of my research, I learned something truly unique about New Mexico that I hadn't previously learned.

As it happens, the official state question in New Mexico is "Red or Green?" If you have never been to New Mexico, this might seem like a strange question, but I assure you it is a very important question that must be answered when ordering the local New Mexican food. This is because most New Mexican cuisine features the *chile* as a staple ingredient, and the most common varieties are the red chile and green chile. Because each is different and their use lends some subtle differences to the foods in which they are included, it is important to discern a diner's preference when ordering. I am a green chile person myself.

This state question gave me a perfect data collection exercise. Before I left to travel to Las Cruces, I walked from door to door in my department and asked ten colleagues this same question: "Red or green"? What I found out was that most people in my department were "green chile people", sharing my preference for green chile. Specifically, eight of them preferred green chile, whereas two preferred red chile. This leads to a simple question: assuming my sample of ten department colleagues forms a reasonable random sample of my university, *what is the proportion of "green chile people" at my university*?

Examples like this are commonly used to teach statistical inference for proportions (e.g., the binomial test). I want to use this example as a concrete testbed to walk through the mechanics of Bayes' theorem. We will enlist some help from JASP along the way.

DEFINING THE UNDERLYING MODEL

To start, we need to figure out what model and associated parameter(s) we will be considering. If we assume that each choice made by my ten colleagues was independent of the others, then we can use the *binomial model*. The binomial model gives the probability of observing a certain number of "successes" out of a set of N independent trials, each of which has only two choices (i.e., red or green). Here, we define a "success" as the choice of green chile. Defining "success" in this way is, of course, completely arbitrary. But as long as we're consistent, it doesn't matter which outcome we define as a success. With the binomial model, we are treating each trial like a coin flip where the probability of success (i.e., the probability of choosing green chile) on each trial is θ. This parameter θ, then, will capture what we are interested in – namely, the proportion of people at my university who are green chile people.

So far, there is nothing Bayesian about this setup. The binomial model is a commonly used model that you may have seen before in other contexts. Its probability distribution is given by

$$p(x \mid \theta) = \frac{N!}{x!(N-x)!} \cdot \theta^x \cdot (1-\theta)^{N-x},$$

where x (our data to observe) represents the number of people who prefer green chiles, and θ (our parameter of interest) represents the population proportion of people who prefer green chiles. Note that in our example, we have $N = 10$ trials and $x = 8$ successes.

DEFINING A PRIOR DISTRIBUTION

What *is* Bayesian about this problem is the thing that we are ultimately interested in. We want to know about the posterior distribution $\pi(\theta \mid \text{data})$, which encodes our belief about the proportion of green chile people θ *after* we have observed the data ($x = 8$ successes out of $N = 10$ trials). Bayes' theorem tells us how to get there – first, we must start by assigning a *prior distribution* on θ. This prior distribution should mathematically encode our *prior* belief about the possible values that could be taken on by θ. How do we do that?

Conceptually, a prior distribution on a parameter θ is simple to imagine. It should encode our prior belief about the values that our parameter θ could take on. Mathematically, it must be a probability distribution, so there are some natural constraints to be considered. For example, as θ represents a probability, it can only range between 0 and 1, and its values must be nonnegative. Thus, our prior distribution $\pi(\theta)$ must be a function that takes on nonnegative values and has its domain between $\theta = 0$ and $\theta = 1$. Further, because it is a probability distribution, the area under the curve between 0 and 1 must be equal to 1. At first glance, it is likely not obvious how to construct such a function. Fortunately, the field of mathematical statistics has provided us with a large library of probability distributions that have these properties. One such distribution is called the *beta distribution*, and it is a particularly nice one to use as our prior for θ (the reasons for this go beyond the scope of this book, but trust me, it works well).

The beta distribution has two positively-valued parameters α and β that control the shape of its probability density function. This means that when we plot the probability density function for $\text{Beta}(\alpha, \beta)$, different values of α and β will result in different overall shapes. Consequently, these different shapes can encode different types of prior belief about the possible values of our binomial model parameter θ. To explain further, consider Figure 2.2, which shows the resulting probability density functions for beta distributions arising from four different combinations of α and β. Note that in the figure, I've chosen some specific values for α and β, but any positive numbers will work.

When we set $\alpha = 1$ and $\beta = 1$, the resulting density function, $\text{Beta}(1,1)$, is flat. This *uniform prior* reflects the prior belief that no one value of θ is more likely than another. Said differently, each value of θ is equally likely. When one does not have strong prior knowledge about the values of θ, this can be a good prior to use. On the other hand, the probability density function for the Beta$(2,2)$ distribution is shaped like a mound with its peak in the center. This prior reflects a belief that θ is most likely to have a value close to $\theta = 0.5$, with decreasing likelihood to take on values that get farther from $\theta = 0.5$

The second row of Figure 2.2 shows two different beta distributions that share some qualities with Beta$(2,2)$, but reflect a stronger prior belief about θ. The Beta$(4,2)$ distribution is still mound shaped, but instead of having its maximum at $\theta = 0.5$, it takes on its maximum at some value *greater* than 0.5. In contrast, the Beta$(2,4)$ distribution takes on its maximum at some value *less* than 0.5. Such distributions would be good to use if we have prior knowledge that indicates that there might be a preference in our population's choices. For example, the Beta$(4,2)$ prior would encode the prior

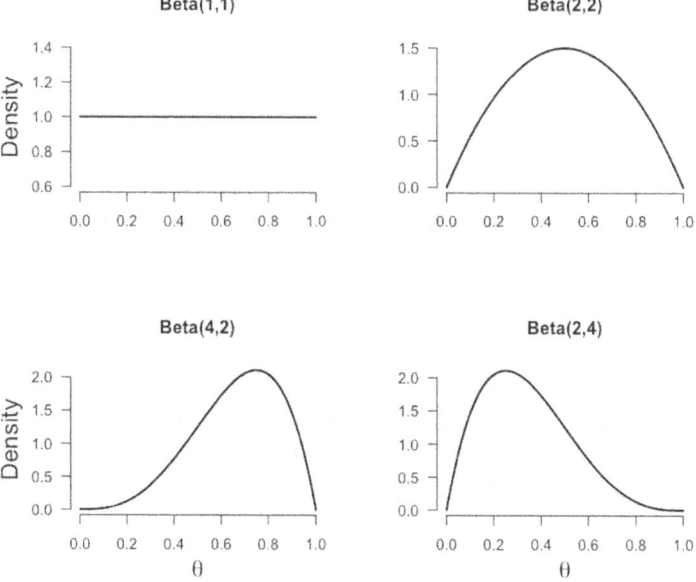

Figure 2.2 Four different beta distributions that arise by varying shape parameters.

belief that θ is greater than 0.5; that is, a prior belief that more than half of my university colleagues prefer green chiles. On the other hand, the Beta$(2,4)$ prior would encode the opposite; namely, a prior belief that *less* than half of my colleagues prefer green chiles.

BAYESIAN UPDATING IN JASP

Once a prior is chosen, the magic of Bayes' theorem is that observing data will mathematically *update* our prior distribution to a *posterior* distribution. This posterior distribution encodes our belief about the possible values of θ *after* data have been observed. So how does this happen? Shortly, we will walk through some of the math behind this Bayesian updating process, but we'll start by using JASP to get a quicker visualization.

Specifically, we will use the "Learn Bayes" module. This module is one of the extra modules that is not loaded by default when the program starts but can be activated as desired. To start the Learn Bayes module, click the "+" icon in the upper right corner of the JASP menu bar. This will open a list of many add-on modules, from which you will select "Learn Bayes". A button for the module will immediately appear in the menu bar. Click that button and select "Binomial Testing".

The Learn Bayes module consists of several sections, and the specific actions that we perform will depend on our goals. For this example, we will use the "Data" section to input our observed data, and we will use the "Hypothesis" section to define the prior distribution on our binomial model parameter θ. To that end, under the "Data" section, we will use the default "Specify counts" option for "Input Type", and under "Count Data", we will input eight successes and two failures. Under the "Hypothesis" section, we'll need to click the "+" icon to add a model – JASP will label this model as "Hypothesis 1", and by default, it will construct a Beta$(1,1)$ (i.e., uniform) prior on θ. For our first example, let's use the Beta$(2,2)$ prior – to do this, simply change each of the numbers in the input fields for α and β from 1 to 2 (see Figure 2.3).

If you're already accustomed to using JASP, you may be somewhat surprised that there isn't much output displayed yet. Don't be alarmed. This module allows us to build exactly what we want to see from our Bayesian updating process. The first thing we want to see

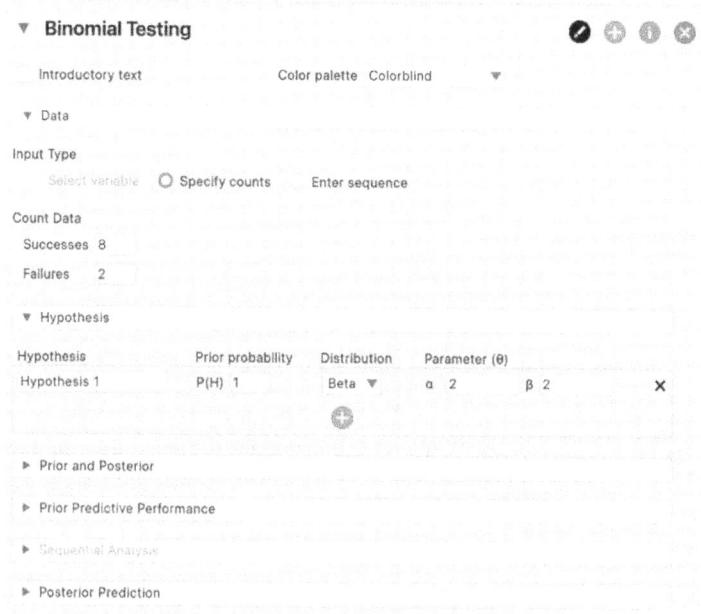

Figure 2.3 A screenshot of the setup screen for Binomial Testing in the JASP Learn Bayes module.

is the posterior distribution. This is easy to build – simply go to the section labeled "Prior and Posterior" and select the option "Prior and posterior distribution". You may ignore the "Type" options below – these only make sense later when we have multiple models under consideration. Go ahead and select the option to display "Observed proportion". This will produce a single plot overlaying both the prior distribution and the resulting posterior distribution, along with an "X" at the observed proportion, $\theta = 8 / 10 = 0.8$ (see Figure 2.4).

In this one figure, we can see Bayes' theorem in action. Before observing any data, the prior distribution on θ reflects a reasonably equivocal belief – namely, that the most likely value for θ is 0.5. This indicates that we have no strong prior leaning toward a preference for green or red chiles in our population. After observing a rather strong preference for green chiles ($\theta = 0.8$) in our sample

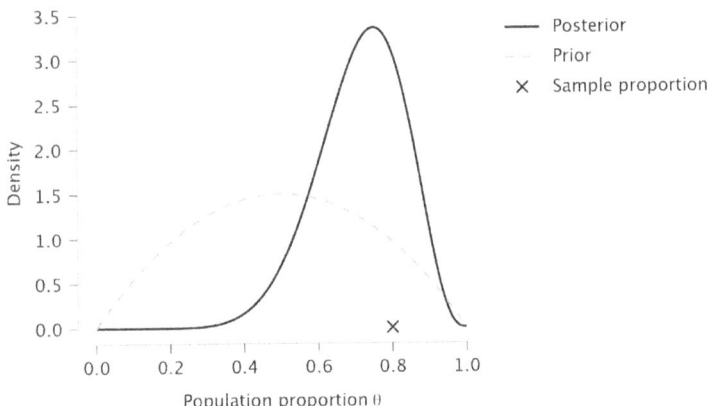

Figure 2.4 The prior and posterior distribution for θ produced in the JASP Learn Bayes module.

of ten, this prior distribution gets squeezed a bit and shifts its peak toward this observed value of θ.

We can go one step beyond this intuitive description, as JASP can give us a bit of detailed information about the posterior distribution. We can display this in the "Prior and Posterior" section by selecting the option "Posterior distribution". Using the default choices under this option will produce Figure 2.5, which gives us (1) a point estimate for our parameter θ, and (2) an interval estimate of θ that incorporates uncertainty. The point estimate is given as the *mean* of the posterior distribution, which in our example is 0.714. Note that other point estimates (e.g., median, mode) can be chosen in the JASP menus. The interval estimate is given as a 95% CI, which ranges from 0.462 to 0.909. Even though the abbreviation looks similar to that of a 95% confidence interval from classical statistics, this is something quite different. Here, the interval displayed by JASP is a 95% *credible interval*, and it represents the central 95% of the posterior distribution. That is, there is a 95% probability that our population proportion θ is between 0.462 and 0.909. In terms of our original research question, this means that we are 95% certain that the proportion of green chile people at my university is between 46.2% and 90.9%.

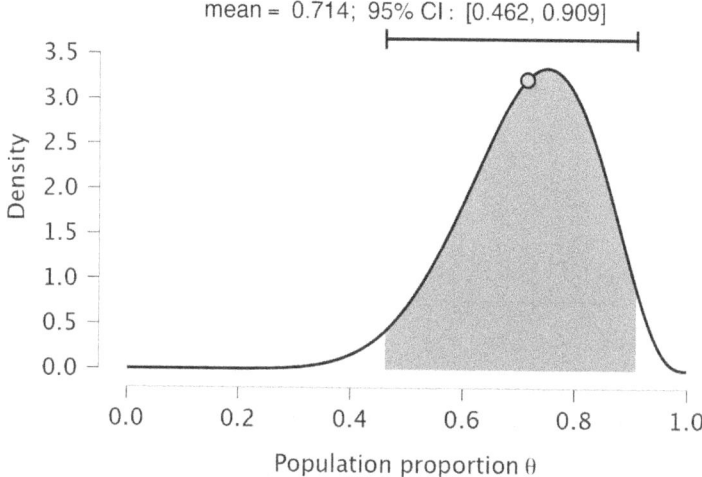

Figure 2.5 Posterior distribution for θ also showing the posterior mean and 95% credible interval.

Compared to the classical confidence interval, I believe this is a much better interval to use for estimating θ, as it captures our measurement uncertainty for θ in terms of a direct probability statement.

THE MATHEMATICS OF BAYESIAN UPDATING

At first glance, the notion that observing data updates the prior distribution to a posterior distribution may seem like magic. However, this magic updating process is nothing more than Bayes' theorem in action. In the next few paragraphs, we'll explore why this is the case. The idea is that we view the posterior distribution as a function of θ, so we simply need to apply Bayes' theorem to every value of θ between 0 and 1. To get a feel for how this works, let's work through the mathematics for one specific value: $\theta = 0.8$ (feel free to do more examples on your own with other values of θ). As we can see in Figure 2.6, the two solid dots represent the density of $\theta = 0.8$ in the prior distribution and the posterior distribution, and the gray arrow indicates the increase in this density that occurs after

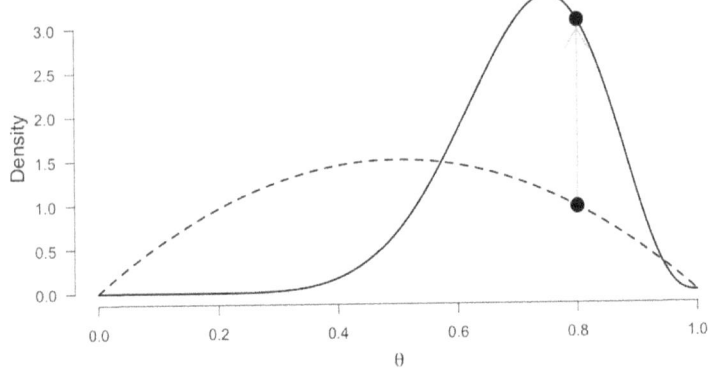

Figure 2.6 A visualization of the Bayesian updating process for $\theta = 0.8$.

observing data. To get the exact values of these densities, we simply need to apply Bayes' theorem at $\theta = 0.8$:

$$\pi\big(\theta = 0.8 \,|\, \text{data}\big) = \pi\big(\theta = 0.8\big) \times \frac{p\big(\text{data} \,|\, \theta = 0.8\big)}{p\big(\text{data}\big)}.$$

This calculation requires four steps. Let's walk through each in turn:

1. First, we need to calculate $\pi\big(\theta = 0.8\big)$. Conceptually, this is the value of the prior distribution at $\theta = 0.8$. As the prior distribution is a $\text{Beta}\big(2,2\big)$ distribution, we need to use the probability density function for the beta distribution. Calculating this by hand is not easy, but computers can do it for us quite easily. In fact, almost any statistical software package can do it, but since this calculation is not crucial to the rest of the book, I'll show you a simple shortcut. You can use the website wolframalpha. com to do it. It works well with natural language input, so just type "Beta(2,2) distribution at x = 0.8", and you'll get the answer 0.96. This aligns well with what we can see in Figure 2.6, as the solid dot on the prior distribution aligns close to 1.0 on the y-axis.

2. Next, we need to calculate $p\big(\text{data} \,|\, \theta = 0.8\big)$. This is the likelihood (under a binomial distribution) of observing $x = 8$

successes if $\theta = 0.8$. To calculate this, we simply need to apply the binomial distribution:

$$p\left(x = 8 \mid \theta = 0.8\right) = \frac{10!}{8!\left(10-8\right)!} \cdot 0.8^8 \cdot \left(1-0.8\right)^{10-8}$$

$$= \frac{10 \cdot 9 \cdot 8 \cdots 2 \cdot 1}{\left(8 \cdots 2 \cdot 1\right)\left(2 \cdot 1\right)} \cdot 0.8^8 \cdot 0.2^2$$

$$= \frac{10 \cdot 9}{2} \cdot 0.1678 \cdot 0.04$$

$$= 0.3020.$$

Note that you could also use Wolfram Alpha to compute this value, but I'll leave that to the interested reader.

3. Last, we need to calculate $p(\text{data})$. Recall from earlier that this is the *marginal probability* of the data, and it represents the weighted average of all the binomial likelihoods (like the one we just calculated in Step 2) over all possible values of θ, weighted by the prior probability of each value of θ. As I indicated earlier, calculating this by hand requires calculus, but JASP will give it to us easily. Let's see how this is done.

In the JASP Learn Bayes module, go to the section labeled "Prior Predictive Performance" and select the "Distribution plot" (the default option of "Conditional" is fine – since we only have one model under consideration right now, this choice doesn't matter). Additionally, select the options to display "Observed number of successes" and "Predictions table". JASP will display Figure 2.7, which represents the marginal distribution $p(\text{data})$ under our Beta$(2,2)$ prior. This distribution represents exactly what we just described – it is the probability of observing each possible outcome (from zero successes up to ten successes), obtained by averaging over all possible values of θ. This distribution is sometimes called the *prior predictive* distribution because it represents the distribution of possible outcomes predicted by our choice of prior. You'll notice that the distribution is similar in shape to our Beta$(2,2)$ prior. You can try for yourself to see how this prior predictive distribution changes by changing the shape parameters of

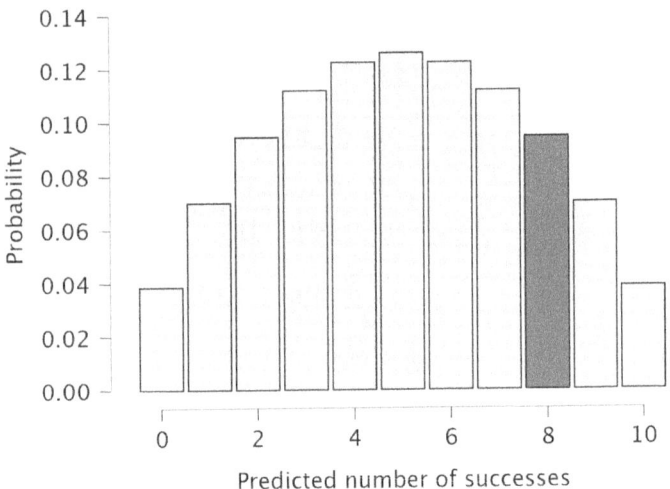

Figure 2.7 Prior predictive distribution from the JASP Learn Bayes module.

the beta distribution in the "Hypothesis" section of the JASP Learn Bayes module.

Along with the distribution, JASP also generates a table that gives us the marginal probabilities of each possible outcome, from $x = 0$ successes to $x = 10$ successes. In that table, we can see that the marginal probability of observing $x = 8$ successes is 0.094. Thus, $p(\text{data}) = 0.094$.

4. With these three components in hand, we can put them all together to calculate the density of $\theta = 0.8$ in the posterior distribution. By Bayes' theorem,

$$\pi(\theta = 0.8 \mid \text{data}) = \pi(\theta = 0.8) \times \frac{p(\text{data} \mid \theta = 0.8)}{p(\text{data})}$$

$$= 0.96 \times \frac{0.3020}{0.094}$$

$$= 3.08.$$

Consulting Figure 2.6, we can see that this result aligns well with our prior and posterior plot, as the solid dot on the posterior is slightly above 3 on the y-axis. Thus, we can see precisely how the y-value *increases* from prior to posterior. This is exactly how the posterior distribution is calculated – at each value of θ between 0 and 1, we just need the prior density at that value of θ and the likelihood of observing $x = 8$ successes at that value of θ. The marginal probability $p(\text{data})$ is fixed. Values of θ which align well with our observation of $x = 8$ successes will be *boosted* in the posterior. In this case, it is values of roughly $\theta = 0.6$ and above that receive this increased support. At the same time, values of θ that do not align well with our observed data will be *suppressed* in the posterior.

BAYESIAN MODEL COMPARISON

As we've just seen, Bayes' theorem uses observed data to mathematically transform a prior distribution on our model parameter(s) into a posterior distribution. This resulting posterior distribution can be used to provide estimates of θ that incorporate measurement uncertainty. The other component of our inferential statistics framework is that of *model comparison*, or *hypothesis testing*. In this final section, we will consider how Bayes' theorem provides us with a useful and coherent framework for model comparison.

In the classical hypothesis testing framework, the typical paradigm is to compare an alternative hypothesis to a null hypothesis. For the context of our chile preference example, this might look like the following:

* a null hypothesis $\mathcal{H}_0 : \theta = 0.5$
* an alternative hypothesis $\mathcal{H}_1 : \theta \neq 0.5$

Under the null hypothesis $\mathcal{H}_0$ we have $\theta = 0.5$, indicating that the proportion of people who prefer green chile is 0.5, or 50%. That is, exactly half prefer green chile, and thus, exactly half prefer red chile. Therefore, this null hypothesis proposes that there is no preference for either red or green chile.

The alternative hypothesis $\mathcal{H}_1$ specifies that θ takes on values not equal to 0.5. That is, under the null hypothesis, there is some preference – if $\theta > 0.5$, this indicates a preference for green chile, whereas if $\theta < 0.5$, this indicates a preference for red chile. Another way to state this alternative hypothesis is that θ is *unconstrained*. Since the Bayesian updating process requires that we place a prior on θ, we typically state the alternative hypothesis more specifically in terms of the prior distribution. That is, we write our two models as:

- $\mathcal{H}_0 : \theta = 0.5$
- $\mathcal{H}_1 : \theta \sim \text{Beta}(2,2)$

This form of the alternative hypothesis is an efficient way to state $\mathcal{H}_1$, as it not only implies that θ is not equal to 0.5 under $\mathcal{H}_1$, but it also indicates exactly how the values of θ are distributed *a priori*.

As we saw in Chapter 1, our goal with model comparison is to see which model does the best job of predicting the data we actually observe. A Bayesian way to do this would be to compare the marginal probability of observing $x = 8$ successes under $\mathcal{H}_1$ (which we found earlier in JASP to be 0.094) to the probability of observing $x = 8$ successes under $\mathcal{H}_0$. Thus, our next step is to introduce the model $\mathcal{H}_0$ in our JASP Learn Bayes module. To do this, go back to the "Hypothesis" section and click the "+" icon to add a new hypothesis. This will add a new hypothesis labeled "Hypothesis 2" – since this will be a null hypothesis, however, let's go with our typical naming convention and change the name to "Hypothesis 0". Under "Distribution", let's change that to a "Spike", which will default to a value of $\theta = 0.5$ (see Figure 2.8).

All the plots that we previously generated in the JASP Learn Bayes module will automatically update to include this new hypothesis $\mathcal{H}_0$. Of particular interest is the prior predictive distribution plot, which now include distributions for both $\mathcal{H}_1$ and $\mathcal{H}_0$ (see Figure 2.9). Recall that this plot shows us the weighted average of the likelihoods of obtaining our observed data of $x = 8$ successes

Hypothesis	Prior probability	Distribution	Parameter (θ)		
Hypothesis 1	P(H) 1	Beta ▼	α 2	β 2	✕
Hypothesis 0	P(H) 1	Spike ▼	θ₀ 0.5		✕

▼ Hypothesis

Figure 2.8 A screenshot from the Hypothesis section of the JASP Learn Bayes module.

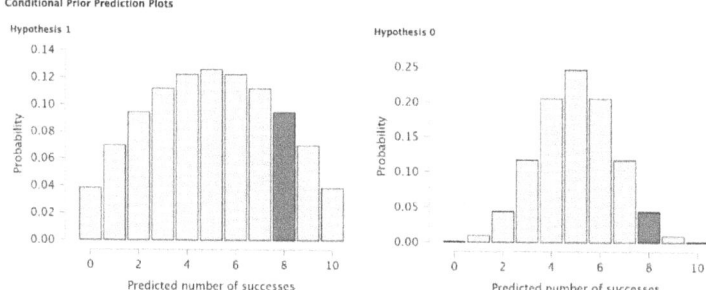

Figure 2.9 Two prior predictive distributions arising from two competing models $\mathcal{H}_1$ and $\mathcal{H}_0$.

over all possible values of θ, weighted by the prior distribution on θ. As we can see in the predictions table that accompanies the plot, the probability of observing eight successes under $\mathcal{H}_1$ is 0.094, whereas the probability of observing eight successes under $\mathcal{H}_0$ is 0.044. That is the observed data are more likely under $\mathcal{H}_1$ than under $\mathcal{H}_0$, making $\mathcal{H}_1$ a better predictor of the actually observed data.

Moreover, we can take the ratio of these two *marginal likelihoods*:

$$\mathrm{BF}_{10} = \frac{p\left(\mathrm{data} \mid \mathcal{H}_1\right)}{p\left(\mathrm{data} \mid \mathcal{H}_0\right)} = \frac{0.094}{0.044} = 2.14.$$

This ratio, which we call the *Bayes factor*, tells us that the observed data are 2.14 times more likely under $\mathcal{H}_1$ than under $\mathcal{H}_0$. We interpret this ratio of predictive performance as *evidence* for $\mathcal{H}_1$ over $\mathcal{H}_0$. In general, observing $\mathrm{BF}_{10} > 1$ indicates that the numerator of this ratio is the larger number, which tells us that $\mathcal{H}_1$ is the better predictor of the observed data. On the other hand, observing $\mathrm{BF}_{10} < 1$ indicates that the denominator is larger, which tells us that $\mathcal{H}_0$ is the better predictor. We can incorporate all this into a concise flowchart (see Figure 2.10), which provides the Bayesian version of our hypothesis testing workflow we introduced in Chapter 1.

One path of this workflow deserves special attention. One of the limitations of the classical hypothesis testing framework we reviewed in Chapter 1 is the inability to directly index support for the null hypothesis $\mathcal{H}_0$. Recall that we can only *reject* $\mathcal{H}_0$ – observing a large

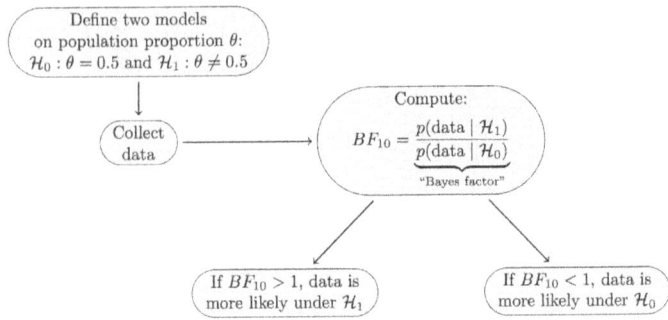

Figure 2.10 A flowchart of the workflow involved in Bayesian hypothesis testing.

p-value doesn't lend any coherent measure of support for $\mathcal{H}_0$ over $\mathcal{H}_1$. This problem goes away when we use Bayesian model comparison. In the event that $\mathcal{H}_0$ is the better predictor of the observed data, we observe a Bayes factor less than 1. Moreover, since the Bayes factor BF_{10} is simply a ratio of predictive performance for $\mathcal{H}_1$ over $\mathcal{H}_0$, we can easily take the reciprocal to recast the ratio as

$$\mathrm{BF}_{01} = \frac{p\left(\text{data} \mid \mathcal{H}_0\right)}{p\left(\text{data} \mid \mathcal{H}_1\right)} = \frac{1}{\mathrm{BF}_{10}}.$$

Mathematically, BF_{01} doesn't tell us anything different from BF_{10}, but conceptually it is a little easier to understand. For example, suppose we observed a Bayes factor of $\mathrm{BF}_{10} = 0.25$. Then taking the reciprocal gives us

$$\mathrm{BF}_{01} = \frac{1}{\mathrm{BF}_{10}} = \frac{1}{0.25} = 4\,.$$

This form can be easily interpreted as "The observed data are 4 times more likely under $\mathcal{H}_0$ than under $\mathcal{H}_1$". To me, this is easier to communicate than the original form, which would be written as "The observed data are 0.25 times as likely under $\mathcal{H}_1$ as under $\mathcal{H}_0$". Both sentences give the same information, but the former is likely to be easier to understand than the latter.

Table 2.1 Evidence categories for Bayes factors

Bayes factor	Evidence category
1–3	Anecdotal evidence
3–10	Moderate evidence
10–30	Strong evidence
30–100	Very strong evidence
>100	Decisive evidence

CATEGORIES OF EVIDENCE

I will conclude this section with a brief discussion about the level of evidence that is provided by a given Bayes factor. Throughout the book, I will adopt the convention of *Jeffreys (1961)*, who proposed the evidence categories in Table 2.1 as indexed by the observed Bayes factor.

It is worth pointing out that one should not worry too much about the specific "break points" proposed in Table 2.1. For example, is a Bayes factor of 9.99 qualitatively different from a Bayes factor of 10.01? The evidence categories are simply a rough benchmark to help when interpreting and communicating the results of a Bayesian analysis. Throughout the remaining chapters, we will refer back to this table often.

PRIORS ON PARAMETERS VERSUS PRIORS ON MODELS

In this final section, we will briefly review what we have learned about Bayesian inference and extend the narrative a bit so that we can discuss a fundamental distinction about what is meant by "prior".

As we saw, Bayes' theorem mathematically expresses one of the most basic ideas in science; namely, that observing data updates our prior belief to a posterior belief. In our previous examples, the subject of our Bayesian updating was the distribution of parameter values θ within a specific model $\mathcal{H}$. In this context, our prior distribution encoded our belief about the parameter values themselves, subject to the specific model under consideration. For example, under $\mathcal{H}_1$, our prior belief about θ was encoded by a Beta$(2,2)$ distribution. Under $\mathcal{H}_0$, our prior belief was encoded by a spike at $\theta = 0.5$.

We can extend our use of Bayes' theorem to the level of the competing models themselves. For any one model $\mathcal{H}$, we can write

$$p(\mathcal{H} \mid \text{data}) = p(\mathcal{H}) \times \frac{p(\text{data} \mid \theta)}{p(\text{data})}.$$

Note that here we use the notation p instead of π because in this context we are interested in the probabilities of the models themselves, not in the probability *distributions* of the parameters under a given model. The distinction is subtle, but important.

Since our goal in Bayesian hypothesis testing is to directly compare the predictive adequacy of our two competing models $\mathcal{H}_0$ and $\mathcal{H}_1$, we can employ Bayes' theorem twice to derive the following:

$$\frac{p(\mathcal{H}_1 \mid \text{data})}{p(\mathcal{H}_0 \mid \text{data})} = \frac{p(\mathcal{H}_1) \times \dfrac{p(\text{data} \mid \mathcal{H}_1)}{p(\text{data})}}{p(\mathcal{H}_0) \times \dfrac{p(\text{data} \mid \mathcal{H}_0)}{p(\text{data})}}.$$

which we can simplify to

$$\frac{p(\mathcal{H}_1 \mid \text{data})}{p(\mathcal{H}_0 \mid \text{data})} = \frac{p(\mathcal{H}_1)}{p(\mathcal{H}_0)} \times \frac{p(\text{data} \mid \mathcal{H}_1)}{p(\text{data} \mid \mathcal{H}_0)},$$

or equivalently,

$$\frac{p(\mathcal{H}_1 \mid \text{data})}{p(\mathcal{H}_0) \mid \text{data})} = \frac{p(\mathcal{H}_1)}{p(\mathcal{H}_0)} \times \text{BF}_{10}.$$

This gives us a very nice way to remember another fundamental tenet of Bayesian hypothesis testing: *posterior odds = prior odds* multiplied by the Bayes factor. Note that the use of the word *odds* has a very specific meaning here. In this case, it means the ratio of probabilities of $\mathcal{H}_1$ and $\mathcal{H}_0$; prior odds refers to the ratio of prior probabilities, and posterior odds refers to the ratio of posterior probabilities. Thus, in addition to serving as an index of the relative predictive performance of our two competing hypotheses, the Bayes factor also indexes the factor by which our relative belief in the two hypotheses can be updated after observing data.

We can leverage this fact in the following way. Suppose that our prior odds in favor of $\mathcal{H}_1$ over $\mathcal{H}_0$ are 1:1 – that is,

$$\frac{p(\mathcal{H}_1)}{p(\mathcal{H}_0)} = 1.$$

This implies that $p(\mathcal{H}_1) = p(\mathcal{H}_0)$. Also, since $\mathcal{H}_0$ and $\mathcal{H}_1$ are complementary events, we have

$$p(\mathcal{H}_1) + p(\mathcal{H}_0) = 1,$$

which can be rewritten as

$$p(\mathcal{H}_0) = 1 - p(\mathcal{H}_1).$$

By substitution, we have

$$p(\mathcal{H}_1) = p(\mathcal{H}_0) = 1 - p(\mathcal{H}_1).$$

Solving this equation gives us a prior probability of $p(\mathcal{H}_1) = 0.5$, and, consequently, $p(\mathcal{H}_0) = 0.5$ also.

Earlier, observing $x = 8$ successes gave us a Bayes factor of $\mathrm{BF}_{10} = 2.14$. In addition to telling us that the observed data are 2.14 times more likely under $\mathcal{H}_1$ than under $\mathcal{H}_0$, it also tells us that the posterior odds for $\mathcal{H}_1$ are 2.14 times the prior odds. Since we started with prior odds of 1:1, this tells us that the posterior odds are now 2.14:1 in favor of $\mathcal{H}_1$.

Just as we did with the prior odds, we can convert these posterior odds directly into a posterior *probability* for $\mathcal{H}_1$. Let's walk through the logic one more time. We'll start by noting that posterior odds equals prior odds multiplied by the Bayes factor; that is,

$$\frac{p(\mathcal{H}_1 \mid \text{data})}{p(\mathcal{H}_0 \mid \text{data})} = 1 \times \mathrm{BF}_{10} = \mathrm{BF}_{10}.$$

Remembering our assumption that $\mathcal{H}_1$ and $\mathcal{H}_0$ are complementary, we can write $p(\mathcal{H}_0 \mid \text{data}) = 1 - p(\mathcal{H}_1 \mid \text{data})$. Substituting directly into the previous expression gives us

$$\frac{p(\mathcal{H}_1 \mid \text{data})}{1 - p(\mathcal{H}_1 \mid \text{data})} = \text{BF}_{10}.$$

We can then solve this equation directly for the posterior probability $p(\mathcal{H}_1 \mid \text{data})$, which gives us

$$p(\mathcal{H}_1 \mid \text{data}) = \frac{\text{BF}_{10}}{1 + \text{BF}_{10}}.$$

Applying this with our observed Bayes factor of $\text{BF}_{10} = 2.14$, we get

$$p(\mathcal{H}_1 \mid \text{data}) = \frac{2.14}{1 + 2.14} = \frac{2.14}{3.14} = 0.682.$$

A similar argument will show that an equivalent expression holds for the posterior probability of $\mathcal{H}_0$ – just use the Bayes factor for $\mathcal{H}_0$ over $\mathcal{H}_1$:

$$p(\mathcal{H}_0 \mid \text{data}) = \frac{\text{BF}_{01}}{1 + \text{BF}_{01}}.$$

In general, this equation works any time our prior odds are set to 1:1. This is a good default position in most hypothesis testing scenarios. On the other hand, if one has reason to have prior odds in favor of one hypothesis over the other, the preceding calculations can be adjusted to reflect these prior odds. This is beyond the scope of the current book, but I invite the interested reader to try to work out the proper equation.

To conclude this section, let us consider what we have done here. When we speak of a "prior" in Bayesian inference, we must specify which type of prior we are considering because there are two types.

1.   For computing a Bayes factor, we must place a prior on the parameter(s) of the underlying model. That is, we must specify

a probability distribution that describes the a priori shape of the distribution of model parameter(s). This prior is then used in the computation of the Bayes factor, specifically as part of the marginal likelihoods that make up the numerator and denominator of the Bayes factor.

2. For computing posterior model probabilities, we must place prior odds on the models under consideration. A common default in simple hypothesis testing situations is 1:1 odds, where the null and alternative hypotheses are considered equally likely, a priori. The Bayes factor can then be used to update these prior odds into posterior odds, which can consequently be used to compute the posterior model probability for either the null or alternative hypothesis.

CHAPTER SUMMARY

In this chapter, we have covered a lot of ground that will let us get started on Bayesian inference. In particular, we encountered the following concepts:

1. We introduced Bayes' theorem and gave a straightforward application to medical testing.
2. We rewrote Bayes' theorem in the context of model parameters and data.
3. We introduced some key vocabulary that will come up throughout the rest of the book, including words like *prior*, *posterior*, and *marginal*.
4. We used JASP to demonstrate Bayesian updating, using an example of a binomial test.
5. We discussed what is meant by a *prior distribution* on model parameters, and worked through the mathematics of how that prior distribution is updated to a *posterior distribution* after observing data.
6. We discovered the *Bayes factor* as a ratio of marginal likelihoods and discussed how it can be used to discern which of two competing models serves as a better predictor of observed data.
7. We introduced Jeffreys' evidence categories as a guide for assessing the evidential value of your observed data.

8. We discussed the difference between placing a prior *distribution* on parameters (which is necessary for computing Bayes factors) and placing prior *odds* on the models/hypotheses themselves. Both are natural things to do in Bayesian inference, but they achieve different goals.

EXERCISES

1. Consider the binomial model from the chapter ($N = 10$ trials, $x = 8$ successes).
    a. In the JASP Learn Stats module, build the following three models in the Hypothesis section:
        i. $\mathcal{H}_1 : \theta \sim \text{Beta}(2,2)$
        ii. $\mathcal{H}_2 : \theta \sim \text{Beta}(2,4)$
        iii. $\mathcal{H}_3 : \theta \sim \text{Beta}(4,2)$
    b. Plot each of the resulting posterior distributions in one figure. An easy way to do this is to go to "Prior and Posterior", choose "Posterior Distribution" + "Joint" + "Overlying".
    c. How are the posteriors similar? How are they different?
    d. Now add a null (spike) model to your list: $\mathcal{H}_0 : \theta = 0.5$. Use the prior predictive distributions to calculate the following Bayes factors for models 1, 2, and 3 against this null model.
    e. Are the Bayes factors similar, or are there differences among them? (*Hint: the point of this exercise is to demonstrate that inference about the posterior distribution doesn't depend too much on the prior used, but Bayes factors do*).
2. In an experiment on ESP (extrasensory perception), participants are required to predict the color (red or black) of a card drawn randomly from a complete deck. Suppose that a set of participants correctly identifies the color on 38 out of 75 trials. Consider a hypothesis test on two models for these data: $\mathcal{H}_0 : \theta = 0.5$ versus $\mathcal{H}_1 : \theta \sim \text{Beta}(1,1)$.
    a. Visually estimate $\pi(\theta = 0.5)$ and $\pi(\theta = 0.5 \mid \text{data})$ from the "Prior and posterior distribution" plot in the JASP Learn Bayes module. Calculate the (estimated) updating factor. Don't think too hard – just eyeball it.

   b.   Calculate the predictive performance of each model against the observed data – that is, $p(\text{data} \mid \mathcal{H}_0)$ and $p(\text{data} \mid \mathcal{H}_1)$. Which model predicted the observed data ($x = 38$ successes) best? How much better was it compared to the other model?

   c.   What do you notice about the answers you obtained in parts (a) and (b)? (*Hint: if you notice that they match, good job . . . you just demonstrated something called the Savage-Dickey density ratio.*)

NOTE

1   Don't worry if you don't consider yourself "mathematically inclined". . . you will still get a lot out of this book!

3

BAYESIAN CORRELATION

In Chapters 1 and 2, we set the stage for our subsequent chapters, each of which feature extensive, guided application of Bayesian methods to some of the most common research designs in the empirical sciences. This chapter will focus on *Bayesian correlation*, and we will revisit two examples from Chapter 1 involving the Big Five personality dataset in the JASP Data Library. Whereas the first two chapters focused on the "what and why" of Bayesian inference, these chapters will largely focus on *how* to implement specific Bayesian analyses. After receiving detailed guidance on how to perform the Bayesian analyses in JASP, you'll also get specific examples of how to report the results, which you can use as a template for your own Bayesian analyses that you'll undoubtedly want to perform later.

WHAT IS *BAYESIAN* CORRELATION?

In Chapter 1, we introduced the Pearson correlation coefficient ρ. Recall that ρ indexes the degree of association between two continuous variables X and Y, and is mathematically defined as

$$\rho = \frac{\sigma_{XY}}{\sigma_X \cdot \sigma_Y},$$

where σ_{XY} is the *covariance* of X and Y, and σ_X and σ_Y are the standard deviations of X and Y, respectively. The correlation coefficient is always constrained between -1 and $+1$, where a correlation of

DOI: 10.4324/9781003470199-3

$\rho = 1$ or $\rho = -1$ represents a perfect positive (or negative, respectively) correlation, and a correlation of $\rho = 0$ would indicate that there is no relationship between X and Y.

At its core, any inference with the Pearson correlation coefficient is concerned with estimating the value of ρ. A *Bayesian* approach to this problem of inference, thus, involves computing the *posterior* distribution for ρ. Recall that Bayes' theorem gives us the following:

$$\pi(\rho \mid \text{data}) = \pi(\rho) \times \frac{p(\text{data} \mid \rho)}{p(\text{data})}.$$

Informally, this equation tells us how to get a posterior distribution for ρ. We first start with a *prior* distribution on ρ, and then update this prior distribution at each value of ρ by a predictive updating factor. Thus, a Bayesian correlation requires us to place a prior on ρ.

Alongside our goal of estimating ρ, a related action that we will perform is a model comparison between a null hypothesis, $\mathcal{H}_0 : \rho = 0$ and an alternative hypothesis $\mathcal{H}_1 : \rho \neq 0$. As we saw in Chapter 2, the Bayesian way to do this is to compare the *prior predictive distributions* at the specific observed data. We do this by taking the ratio:

$$\text{BF}_{10} = \frac{p(\text{data} \mid \mathcal{H}_1)}{p(\text{data} \mid \mathcal{H}_0)}.$$

This ratio, which we call the *Bayes factor* for $\mathcal{H}_1$ over $\mathcal{H}_0$, gives us the relative predictive adequacy of $\mathcal{H}_1$ over $\mathcal{H}_0$. If $\text{BF}_{10} > 1$, this tells us that the numerator $p(\text{data} \mid \mathcal{H}_1)$ is greater than the denominator $p(\text{data} \mid \mathcal{H}_0)$. That is, $\mathcal{H}_1$ does a better job of predicting the observed data than $\mathcal{H}_0$. On the other hand, if $\text{BF}_{10} < 1$, this tells us that the numerator $p(\text{data} \mid \mathcal{H}_1)$ is *less* than the denominator $p(\text{data} \mid \mathcal{H}_0)$. That is, $\mathcal{H}_0$ does a better job of predicting the observed data than $\mathcal{H}_1$. Moreover, beyond telling us which model is the better predictor of the observed data, the Bayes factor quantifies *by what factor* the observed data are more likely under the winning model. In this way, BF_{10} is a measure of the *evidence* for the winning model.

DEFINING A PRIOR FOR ρ

Importantly, both actions we just discussed (estimating ρ and comparing models about ρ) require us to place a prior on ρ. This prior gives us the ability to mathematically specify our a priori uncertainty about ρ before observing data. In Chapter 2, we saw our first example of a prior distribution for the binomial parameter θ. Because θ is a probability, it necessarily takes on values between 0 and 1, making the beta distribution a useful tool for specifying this uncertainty. But here, the situation is a little different. The correlation coefficient ρ takes on possible values extending from -1 to 1, so the beta distribution we used in Chapter 2 will not work – at least without some modification.

Based on this foreshadowing, you might not be surprised that there is a way to modify the beta distribution so that its domain of values "stretches" to fit the range of expected values for the correlation. This *stretched beta distribution* is described in detail by *Ly, Marsman, and Wagenmakers (2017)*, and it is the prior that JASP uses in its Bayesian correlation procedure. While the technical details of this prior are beyond the scope of this book, I can describe it intuitively here. Basically, the stretched beta prior starts off as a symmetric beta distribution, so it is of the form $\text{Beta}(a,a)$ for some shape parameter a. Because the domain of the beta distribution is the interval $[0,1]$, we must transform the distribution to have domain $[-1,1]$ in order to serve as a reasonable prior for a correlation coefficient. In other words, we must "stretch" it to cover the wider interval. The shape of the distribution is then controlled by a single parameter, called the *width*, which is equal to $\kappa = 1/a$. Figure 3.1 shows three examples of a stretched beta distribution, each defined by its specific value of the width parameter κ:

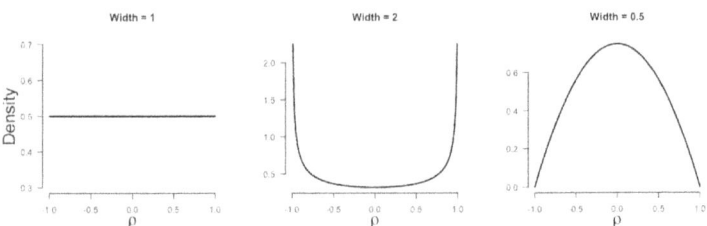

Figure 3.1 Three different stretched beta distributions.

In the leftmost plot of Figure 3.1, we see a stretched beta distribution with width $\kappa = 1$. This is a *uniform* distribution, and it encodes the prior belief that all values of ρ are equally likely before observing data. This uniform prior is the default prior specified in the JASP Bayesian correlation test, which we will see shortly. The middle plot of Figure 3.1 shows a stretched beta distribution with width $\kappa = 2$. Notice that this plot takes on its minimum value at $\rho = 0$ and increases as ρ approaches ± 1. This prior encodes a different type of prior belief – namely, an expectation that nonzero values of ρ are more likely than $\rho = 0$. Finally, the rightmost plot of Figure 3.1 shows a stretched beta distribution with width $\kappa = 0.5$. Compared to the previous plot, this distribution specifies exactly the opposite type of prior belief, namely that $\rho = 0$ is the most likely value for ρ, and nonzero values of ρ become less likely as ρ approaches ± 1.

At this point, you may be asking, "Does it matter how I choose this prior?" The answer is "yes". As we know from one of the exercises in Chapter 2, the specific choice of prior can affect the value of the Bayes factor we observe, so clearly the choice of prior does have some nontrivial impact on our inference. But, as we'll learn in the next section, JASP provides us with a way to be transparent about our choice of prior and simultaneously assess the sensitivity of our conclusions to that specific choice of prior.

EXAMPLE 1: THE ASSOCIATION BETWEEN NEUROTICISM AND AGREEABLENESS

Let us now walk through a detailed example of performing a Bayesian correlation in JASP. We'll use the "Big Five Personality Traits" dataset from the JASP Data Library (see Chapter 1). Recall that these data come from *Dolan et al.'s (2009)* Dutch version of the NEO PI-R that was administered to 500 first-year psychology students. Each row represents one of the students, and the columns correspond to the composite scores (scaled from 1 to 5) for each student on each of the "Big Five" personality dimensions: Agreeableness, Conscientiousness, Neuroticism, Extraversion, and Openness to Experience.

Once you load the dataset (remember, you want the dataset whose icon does not have the JASP logo), click on the "Regression" button in the top menu bar. From here, choose "Correlation" from the "Bayesian" section. Since we will be testing the association between a couple of pairs, let's go ahead and move all five variables into the analysis box.

Immediately after doing this, you'll see a matrix of correlations appear in the results page on the right-hand side of the screen (along with their Bayes factors BF_{10}). Don't worry about this for now, as we will be paying closer attention to some individual pairs. To this end, click the "Plot Individual Pairs" button. Here, you will specify the pair of interest; namely, Neuroticism and Agreeableness. Go ahead and highlight those variable names and click the right arrow button to move them into the box of pairs. Immediately, you will see a scatterplot appear, as that option is checked by default. You'll also want to select the "Prior and posterior" plot. Figure 3.2 shows the JASP analysis screen with all these options selected.

The options we've selected will produce two plots of interest. We've already seen one of these in Chapter 1 – the scatterplot, which shows a negative relationship between Neuroticism and

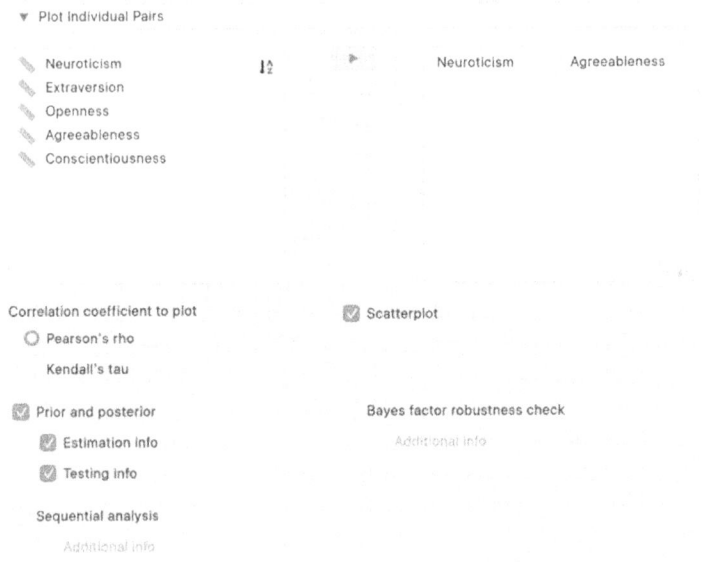

Figure 3.2 A JASP screenshot showing part of the analysis screen for a Bayesian correlation.

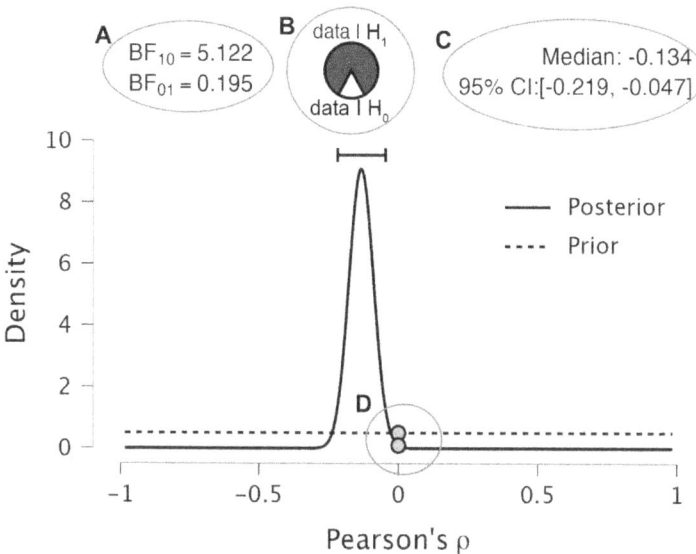

Figure 3.3 A JASP prior and posterior plot for the Bayesian correlation between neuroticism and agreeableness.

Agreeableness. The other plot is new for us and will be worth going through in detail. This "Prior and posterior" plot is packed full of information, so much so that a simple figure caption will not suffice. Thus, I have annotated several different components of Figure 3.3 to describe in detail, which now follows:

Annotation A: The first thing you'll notice in the upper left of the prior and posterior plot is a pair of Bayes factors; $BF_{10} = 5.122$ and $BF_{01} = 0.195$. Both numbers give exactly the same information – just in a different way. Remember that BF_{10} represents the relative likelihood of the observed data under $\mathcal{H}_1$ compared to $\mathcal{H}_0$. Thus, $BF_{10} = 5.122$ means that the data are 5.122 times more likely under the alternative hypothesis $\mathcal{H}_1$ than under the null hypothesis $\mathcal{H}_0$. On the other hand, BF_{01} is the opposite – namely, the relative likelihood of the observed data under $\mathcal{H}_0$ compared to $\mathcal{H}_1$. The two Bayes factors are *reciprocals* of each other. That is,

$$\mathrm{BF}_{01} = \frac{1}{\mathrm{BF}_{10}} = \frac{1}{5.122} = 0.195 \,.$$

Of course, you may wonder which you should report. The choice is yours, but my advice is to report the one that is larger than 1. In this case, we would simply communicate that the observed data are 5.122 times more likely under $\mathcal{H}_1$ than under $\mathcal{H}_0$.

Annotation B: In the top center of the prior and posterior plot is a "pizza plot",[1] which is a circle graph (i.e., a pie chart) that gives a nice visual representation of the relative likelihood of the observed data under both competing models. In this case, we can see that the portion representing the likelihood of the data under $\mathcal{H}_1$ is quite a bit larger than the portion for $\mathcal{H}_0$. In fact, the red sector has exactly 5.122 times more area than the white sector.

Annotation C: Remember that Bayes' theorem gives us a way to compute the posterior distribution for the parameter of interest in our analysis. In this case, that parameter is the population correlation coefficient ρ, and the upper right corner of the prior and posterior plot gives some important summary statistics for the posterior distribution of ρ. The two things that JASP outputs are the *median* of the posterior distribution and the 95% credible interval for ρ. Here, we see that the median posterior value for ρ is -0.134. Notice that this matches the observed correlation in the correlation matrix. However, keep in mind that these two numbers, while they match, are conceptually *different*. The one in the correlation matrix is the value of the correlation coefficient that is computed from the *sample* — that is, $r = -0.134$. On the other hand, the one in the prior and posterior plot is the model-estimated value of the population correlation coefficient ρ — in this case, given as the median value of the posterior distribution for ρ. Alongside this median, we get a 95% credible interval of $[-0.219, -0.047]$, which means that there is a 95% probability that the population correlation coefficient ρ lies between -0.219 and -0.047.

Annotation D: Finally, let's focus on the plot itself. Here, you see the prior distribution as a dashed black line — notice that it is a uniform distribution over the interval $[-1,1]$, which is the default prior specification in JASP (i.e., a stretched beta prior with width $\kappa = 1$). After observing data, this prior distribution is updated (via Bayes' theorem) to the posterior distribution, which is given as a solid black line. We already discussed some summary statistics of the posterior distribution when we talked about Annotation C (namely, the median and the 95% credible interval). Here, I want to focus on the two "dots" that are circled as Annotation D. These two dots are placed there because they represent the density (i.e., height) of $\rho = 0$, which is the value that ρ is restricted to in the null hypothesis $\mathcal{H}_0$.

Specifically, we can see that the probability density of $\rho = 0$ in the posterior distribution is *less* than the density of $\rho = 0$ in the prior distribution. Said differently, our belief that $\rho = 0$ (i.e., our belief in $\mathcal{H}_0$) *decreases* after observing data. Remarkably, the factor by which this belief decreases is exactly the Bayes factor $BF_{10} = 5.122$. This fact (which we foreshadowed in Exercise 2 of Chapter 2) is known as the *Savage-Dickey density ratio* (e.g., *Wagenmakers, Lodewyckx, Kuriyal, & Grasman, 2010*), and it represents an efficient method for approximating Bayes factors in situations where one can only estimate posterior distributions. Here, that is not so important. What *is* important here is that it is a visually intuitive representation of the Bayesian updating process, showing how our belief in a specific model ($\mathcal{H}_0$) decreases after observing data.

COMPUTING THE POSTERIOR MODEL PROBABILITY

We just saw that our observed correlation of $r = -0.134$ lends moderate evidence to $\mathcal{H}_1$, which states that there is a nonzero correlation between Neuroticism and Agreeableness. This evidence for $\mathcal{H}_1$ is indexed by a Bayes factor of $BF_{10} = 5.122$. We saw in Chapter 2 that this Bayes factor can be interpreted in two ways: (1) as relative predictive adequacy, and (2) as an updating factor for prior model odds. Interpreting as relative predictive adequacy, we see that the observed data are 5.122 times more likely under $\mathcal{H}_1$ than under $\mathcal{H}_0$. This is the interpretation that we already described earlier.

Interpreting as an updating factor for prior model odds gives us the ability to compute a posterior model probability. Remember that this is different from the posterior *distribution* on ρ, which quantifies our uncertainty about the values that ρ can take on. In contrast, the posterior model probability is a single number that reflects how likely $\mathcal{H}_1$ (or $\mathcal{H}_0$) is to be the correct model. We can get this posterior model probability by considering that the Bayes factor is an updating factor for model odds. If we begin with prior model odds of 1:1 (i.e., we believe that $\mathcal{H}_1$ and $\mathcal{H}_0$ are equally likely, a priori), then the posterior model odds are obtained by multiplying these odds by the Bayes factor. This gives posterior model odds of 5.122:1 in favor of $\mathcal{H}_1$. As we saw in Chapter 2, we can easily convert this statement about odds into a probability:

$$p\left(\mathcal{H}_1 \mid \text{data}\right) = \frac{\text{BF}_{10}}{1 + \text{BF}_{10}} = \frac{5.122}{1 + 5.122} = \frac{5.122}{6.122} = 0.837.$$

Thus, the posterior model probability of $\mathcal{H}_1$ is equal to 0.837. Remember, the posterior odds and posterior probability are just two different ways of communicating the same information; namely, that our observed data indicate a reasonably large preference for $\mathcal{H}_1$ over $\mathcal{H}_0$.

WHAT IF WE CHOOSE A DIFFERENT PRIOR ON ρ?

Certainly, we have demonstrated that the preceding analysis presents us with a lot of information to guide our inference about the possible association between Neuroticism and Agreeableness. First, we have positive support for $\mathcal{H}_1$ in the form of a Bayes factor $\text{BF}_{10} = 5.122$, which tells us that the observed data (a sample correlation of $r = -0.134$ with a sample of $N = 500$) is 5.122 times more likely under $\mathcal{H}_1$ than under $\mathcal{H}_0$. Further, given that we've chosen $\mathcal{H}_1$ as our winning model, it makes sense to *estimate* the value of the population correlation coefficient ρ. This estimate comes in the form of a posterior distribution for ρ, from which we get a median value of $\rho = -0.134$ and a 95% credible interval of $\left[-0.219, -0.047\right]$. Thus, we can reasonably conclude that there is a meaningful negative association between Neuroticism and Agreeableness.

However, basing our conclusions on this one analysis could potentially be misleading. As we know from Exercise 1 in Chapter 2, the specific value of the Bayes factor we obtain from our observed data depends on the specific prior we place on the model parameter. Could it be the case that our conclusions are based on a pattern of results that depend solely on the prior we chose? What would happen if we used a different prior?

Fortunately, this is a question that can be easily addressed in JASP. In the analysis window, we can change the width κ of the stretched beta prior and immediately see the impact on our inference in the results window. For example, it is easy to generate the two prior and posterior plots below by simply changing the prior width to 0.5 (left plot of Figure 3.4) and 2.0 (right plot of Figure 3.4). These two values represent two reasonable extremes for the shape of the stretched beta prior – for $\kappa = 0.5$, more prior mass is placed on small (absolute) values of ρ, whereas for $\kappa = 2.0$, more prior mass is placed on large (absolute) values of ρ. Two things are immediately clear from Figure 3.4. First, $\mathcal{H}_1$ is preferred in both cases, though the size of the observed Bayes factor does change. Second, the estimates of the posterior distributions are virtually identical. This gives further support to the claim we formulated in Exercise 1 of Chapter 2 – namely, that the choice of prior matters little for estimation, but it matters a lot for model comparison.

While one can manually change values of the stretched beta prior width κ, there is a better solution already implemented in JASP. In the analysis window under the section "Plot Individual Pairs", you can check the option for "Bayes factor robustness check", which will produce Figure 3.5. This plot displays the observed Bayes factors obtained over a range of different settings for the stretched beta

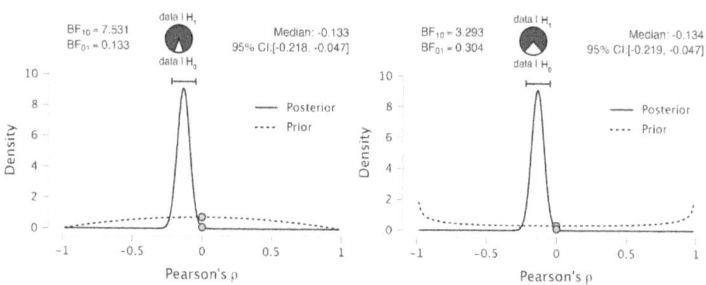

Figure 3.4 Prior and posterior plots for two different widths κ of the stretched Beta prior. The left plot represents $\kappa = 0.5$, and the right plot represents $\kappa = 2.0$.

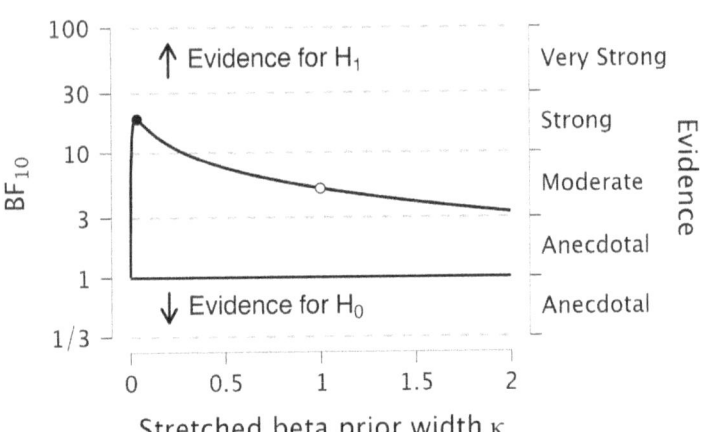

Figure 3.5 A sensitivity analysis for the Bayesian correlation between Neuroticism and Agreeableness, which shows the range of Bayes factors obtained over a range of possible widths κ for the stretched Beta prior.

prior width κ. More generally, such a *sensitivity analysis* allows a user to see exactly how the Bayes factor depends on the shape of the prior. Here, we can see that small values of κ (where the prior is more peaked around $\rho = 0$) give larger Bayes factors, whereas larger values of κ (where the prior peaks at the ends) give smaller Bayes factors. Why would that be?

This is reasonably easy to explain, at least conceptually. When we use a small value of κ, we are saying that our prior belief is that the correlation ρ should most likely take on small values, with the most likely value being $\rho = 0$. On the other hand, when we use a large value of κ, our prior belief is that the correlation ρ should most likely take on *large* values, with $\rho = 0$ being the *least likely* possibility. After observing data, this prior is updated to the posterior distribution, which we've already seen is similar for both small values of κ (e.g., $\kappa = 0.5$) and large values of κ (e.g., $\kappa = 2$). Now consider the size of the prior-to-posterior updating that must occur at $\rho = 0$. When we start with a small value of κ, the factor by which the prior density of $\rho = 0$ changes is much larger than the case where we use a large value of κ. Recalling our discussion earlier about the Savage-Dickey density ratio, we then realize that this updating factor at $\rho = 0$ *is* the Bayes factor BF_{10}. Thus, our Bayes factors generally should *decrease* as the stretched beta prior width increases. As we can see in Figure 3.5, this is basically what we see. Broadly, it is important to note that across a wide range of reasonable prior settings, we obtain moderate evidence for $\mathcal{H}_1$.

HOW TO COMMUNICATE THE RESULTS

We'll wrap up this example with a brief demonstration of how to communicate the results of this Bayesian analysis. There are three main components that we should include:

1. Model definitions
2. Model comparison and parameter estimation
3. Sensitivity analysis.

Model definitions: We should always begin with a definition of the models being considered. This is no different from what we would do in classical inference, but in our Bayesian context, we must additionally specify the prior distribution under $\mathcal{H}_1$. Here is an example of what this could look like for the analysis we just performed:

> *Under the null hypothesis, we expect a correlation of 0 between Neuroticism and Agreeableness. Thus, we define $\mathcal{H}_0 : \rho = 0$. The alternative hypothesis is $\mathcal{H}_1 : \rho \neq 0$, and we took the prior distribution for ρ under $\mathcal{H}_1$ to be a stretched beta distribution with width $\kappa = 1.0$. This distribution is the default prior used in JASP, and it assigns a uniform prior probability to all values of ρ between -1 and $+1$.*

Model comparison and parameter estimation: Next, we should report the results of our model comparison and parameter estimation. Note that some authors suggest that estimation should be done first, but I disagree. The reason is that estimation of the posterior distribution only makes sense if we know that $\rho \neq 0$, which is only the case if $\mathcal{H}_1$ is the "winning" model. Thus, my suggestion is to report the model comparison *first*, then if warranted, proceed with the parameter estimation from the posterior distribution. Here is an example from our analysis:

> *We found a Bayes factor of* $\mathrm{BF}_{10} = 5.122$, *which means that the observed data are approximately 5 times more likely under* $\mathcal{H}_1$ *than under* $\mathcal{H}_0$. *Using the suggested evidence categories from Jeffreys (1961), this result indicates moderate evidence in favor of* $\mathcal{H}_1$. *Further, the posterior distribution for* ρ *had a median of* $\rho = -0.134$, *with a 95% credible interval that ranges from* -0.219 *to* -0.047.

We can also report our posterior model probability:

> *Further, as the Bayes factor represents the updating factor on prior model odds, we can translate it directly into a posterior model probability. Indeed, if we begin with 1:1 prior odds for* $\mathcal{H}_1$ *and* $\mathcal{H}_0$ *(i.e., assuming* $\mathcal{H}_1$ *and* $\mathcal{H}_0$ *are equally likely, a priori), the posterior odds are* $5.122:1$ *in favor of* $\mathcal{H}_1$. *This is equivalent to a posterior model probability of* $p(\mathcal{H}_1 \mid \text{data}) = 0.837$.

Sensitivity analysis: Finally, we need to communicate how our inference depends on our specific choice of prior distribution for the model parameter(s). An example of this might look like the following:

> *Since Bayes factors depend on the specific choice of prior distribution for our model parameter* ρ, *we additionally performed a sensitivity analysis to assess the impact of this prior choice on our inference. In Figure 3.5, we have plotted the Bayes factor as a function of the width* κ, *which is the parameter that governs the shape of our prior distribution. We can see that across a wide range of reasonable values for this shape parameter, the obtained Bayes factor represents moderate evidence in favor of* $\mathcal{H}_1$.

EXAMPLE 2: THE ASSOCIATION BETWEEN EXTRAVERSION AND CONSCIENTIOUSNESS

For our second example, we will explore the association between Extraversion and Conscientiousness. Recall from our previous work in Chapter 1 that this pair exhibited a nonsignificant correlation, though our methods of classical inference based on p-values did not allow us to conclude any sort of positive support for $\mathcal{H}_0$. Let's see what a Bayesian analysis will let us do.

Working again in the "Plot Individual Pairs" section, simply highlight the variable names `Extraversion` and `Conscientiousness` and move them each into the box of pairs. Immediately, you will see a scatterplot appear, just like in the previous example. Additionally, go ahead and select the boxes for "Prior and Posterior" and "Bayes factor robustness check". Let's first consider the prior and posterior plot, shown in Figure 3.6.

Just like in Example 1, displayed in the upper left of the prior and posterior plot is a pair of Bayes factors; $BF_{10} = 0.158$ and $BF_{01} = 6.329$. In this case, we see that we have positive support for the null hypothesis $\mathcal{H}_0$, as the Bayes factor tells us that the observed correlation ($r = 0.065$) is 6.329 times more likely under $\mathcal{H}_0$ than under $\mathcal{H}_1$. This information is also conveyed visually in the pizza plot, where the area corresponding to the likelihood of the data under $\mathcal{H}_0$ is quite a bit larger than that corresponding to the likelihood of the data under $\mathcal{H}_1$.

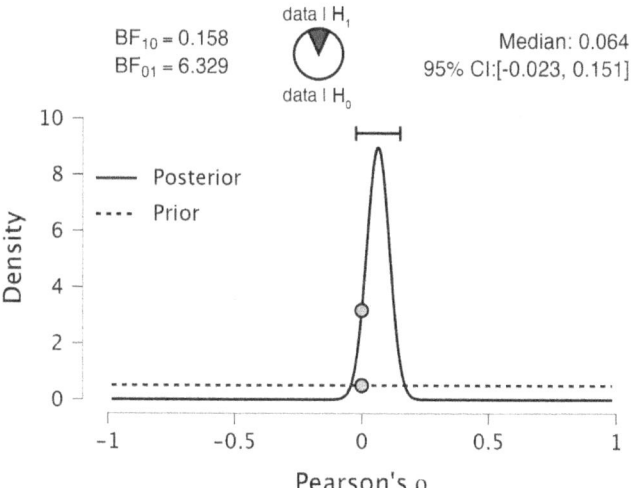

Figure 3.6 A JASP prior and posterior plot for the Bayesian correlation between Extraversion and Conscientiousness.

Next, let's look at the actual plots of the prior and posterior distributions for ρ. Again, the default prior for the population correlation coefficient ρ is a stretched beta distribution with width $\kappa = 1$, which is equivalent to a uniform distribution over the interval $[-1,1]$. After observing data, the posterior is now much more peaked over the area close to $\rho = 0$. In fact, the dots on the prior and posterior plot show that the probability density associated with the specific value $\rho = 0$ *increases* after observing data – in fact, recall that the magnitude of this increase is given by the Bayes factor, so we see that this probability increases by a factor of 6.329. In all, our Bayesian analysis has increased our confidence that the population correlation coefficient is $\rho = 0$, giving us positive evidence for $\mathcal{H}_0$.

Even though it is not included in JASP, we can easily convert this Bayes factor into a posterior model probability for the "winning" model $\mathcal{H}_0$. Indeed, if we assume 1:1 model odds for $\mathcal{H}_1$ and $\mathcal{H}_0$, the Bayes factor represents the updating factor to give us posterior model odds. Thus, the posterior model odds are 6.329:1 in favor of $\mathcal{H}_0$. A little arithmetic allows us to convert these posterior odds to a posterior model probability:

$$p(\mathcal{H}_1 \mid \text{data}) = \frac{\text{BF}_{01}}{1 + \text{BF}_{01}} = \frac{6.329}{1 + 6.329} = \frac{6.329}{7.329} = 0.864.$$

Let's also consider the posterior estimates given in the upper right corner of the prior and posterior plot. It is important to note that estimating the posterior distribution only makes sense if we assume $\mathcal{H}_1$ is the correct model (i.e., we *condition on* $\mathcal{H}_1$). Since the null hypothesis $\mathcal{H}_0$ is just a point $\rho = 0$, conditioning on $\mathcal{H}_0$ leaves no role for any sort of estimation – in such a case, there is no variability or uncertainty in the values of ρ. Nonetheless, the standard JASP output is to provide these posterior estimates. I would argue that this is a good feature, though, as even if we were to condition on $\mathcal{H}_1$, we can see that the 95% credible interval $[-0.023, 0.151]$ contains $\rho = 0$. This gives us yet another way to index positive support for the *absence* of a correlation between extraversion and conscientiousness.

Finally, we should check the extent to which our obtained Bayes factor depends on the specific prior we placed on ρ under $\mathcal{H}_1$. The Bayes factor robustness plot in Figure 3.7 displays the observed

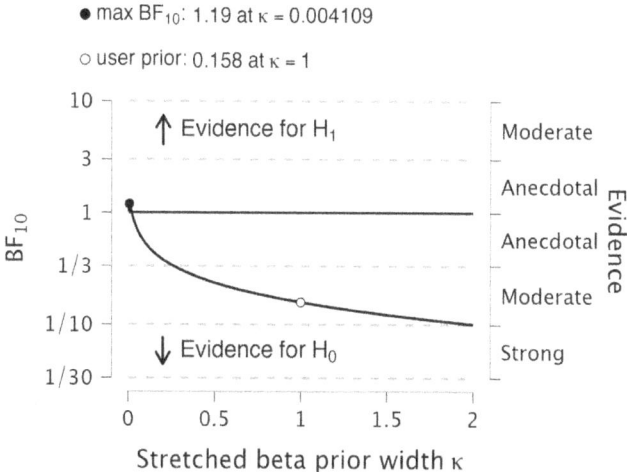

Figure 3.7 A sensitivity analysis for the Bayesian correlation between Extraversion and Conscientiousness, showing the range of Bayes factors obtained over a range of possible widths κ for the stretched beta prior.

Bayes factors obtained over a range of different settings for the stretched beta prior width κ. From Figure 3.7, we see that across a wide range of reasonable settings for the stretched beta prior width κ, we obtain moderate evidence for $\mathcal{H}_0$.

HOW TO COMMUNICATE THE RESULTS

As before, we'll wrap up this example with a template to use for communicating the results of this Bayesian analysis. As you read this, look for the places where our writeup is similar to Example 1, and perhaps more importantly, look for where it is different.

Model definitions: Here, we used the same model definitions from Example 1:

Under the null hypothesis, we expect a correlation of 0 between Extraversion and Conscientiousness. Thus, we define $\mathcal{H}_0 : \rho = 0$. The alternative hypothesis is $\mathcal{H}_1 : \rho \neq 0$, and we took the prior distribution for ρ under $\mathcal{H}_1$ to be a stretched beta distribution with width $\kappa = 1.0$.

This distribution is the default prior used in JASP, and it assigns a uniform prior probability to all values of ρ between −1 and +1.

Model comparison and parameter estimation: As before, we report the model comparison *first*, then only if warranted, proceed with the parameter estimation from the posterior distribution. In this case, since $\mathcal{H}_0$ is our winning model, we can omit the information about the posterior distribution:

We found a Bayes factor of $\mathrm{BF}_{01} = 6.329$, which means that the observed data are approximately 6.3 times more likely under $\mathcal{H}_0$ than under $\mathcal{H}_1$. Using the suggested evidence categories from Jeffreys (1961), this result indicates moderate evidence in favor of $\mathcal{H}_0$, indicating the absence of a correlation between Extraversion and Conscientiousness.

Further, as the Bayes factor represents the updating factor on prior model odds, we can translate it directly into a posterior model probability. Indeed, if we begin with 1:1 prior odds for $\mathcal{H}_1$ and $\mathcal{H}_0$ (i.e., assuming $\mathcal{H}_1$ and $\mathcal{H}_0$ are equally likely, a priori), the posterior odds are 6.329:1 in favor of $\mathcal{H}_0$. This is equivalent to a posterior model probability of $p(\mathcal{H}_0 \mid \mathrm{data}) = 0.864$.

Sensitivity analysis: This example will be largely the same as Example 1, except this time our evidence is for $\mathcal{H}_0$ rather than $\mathcal{H}_1$:

Since Bayes factors depend on the specific choice of prior distribution for our model parameter ρ, we additionally performed a sensitivity analysis to assess the impact of this prior choice on our inference. In Figure 3.7, we have plotted the Bayes factor as a function of the width κ, which is the parameter that governs the shape of our prior distribution. We can see that across a wide range of reasonable values for this shape parameter, the obtained Bayes factor represents moderate evidence in favor of $\mathcal{H}_0$.

CHAPTER SUMMARY

Let's recap what we've learned in this chapter:

1. We introduced the Bayesian correlation test, which compares two models about the population correlation coefficient ρ: a

null hypothesis $\mathcal{H}_0$, where $\rho = 0$, and an alternative hypothesis $\mathcal{H}_1$, where ρ ranges from -1 to 1 according to a prior distribution on ρ.

2. We discussed how to use the *stretched beta distribution* as a default prior for the population correlation coefficient ρ. This prior has one *shape* parameter κ, which is called the "width" of the prior.

3. We worked through two detailed examples in JASP and gave examples of how to communicate the results of our Bayesian analyses.

4. Because Bayes factors depend on the choice of prior for the model parameter(s), we discussed the need for sensitivity analyses. In JASP, this is easily obtained using the "Bayes factor robustness check", which plots the obtained Bayes factor as a function of the prior's shape parameter.

EXERCISES

1. In this exercise, you will continue to use the Big Five Personality Traits dataset in JASP to perform a Bayesian correlation.

   a. Construct a scatterplot showing the relationship between Openness and Agreeableness. What do you notice?

   b. What is the default prior used by JASP for the Bayesian correlation test? What does it imply for our prior belief about the (population) correlation coefficient ρ?

   c. Select "Prior and posterior" in the "Plot Individual Pairs" menu. Which model ($\mathcal{H}_1$ or $\mathcal{H}_0$) best predicts the observed correlation? How do you know?

   d. Report the median posterior value and 95% credible interval for ρ.

   e. Suppose you include this analysis in a paper you are trying to publish. A reviewer comments "I reject your Bayesian analysis because the results depend too much on the choice of prior." How do you respond?

2. In this exercise, you are going to use a great tool for doing a *secondary* Bayesian analysis on published results. It is called the "Summary Statistics" module and is accessible from the list of add-on modules in JASP. Suppose you are reading a published

paper and see that the authors obtain a sample of size 175 and find a significant correlation of $r = 0.15$, $p = 0.048$.

a.   Open the "Summary Statistics" module, select "Bayesian correlation", and input this sample size and observed correlation in the module. You'll immediately get both a p-value and a Bayes factor. Interpret this result as you would in classical inference (i.e., based on the p-value). Which model ($\mathcal{H}_0$ or $\mathcal{H}_1$ is supported) by this result?

b.   Interpret this result as you would in Bayesian inference (i.e., based on the Bayes factor). Which model is supported from this analysis?

c.   What do you think is going on here? (Hint: use a search engine or generative AI chatbot to explore "Lindley's paradox").

NOTE

1   E.J. Wagenmakers gives a fun way of thinking about pizza plots in a post on his *Bayesian Spectacles* blog: https://www.bayesianspectacles.org/lets-poke-a-pizza-a-new-cartoon-to-explain-the-strength-of-evidence-in-a-bayes-factor/.

THE BAYESIAN *t*-TEST

In the previous chapter, we walked through detailed examples of performing a Bayesian correlation. This chapter will be very similar in many respects – indeed, the mechanics of performing many of the different Bayesian tests that JASP provides are similar enough that a little bit of knowledge about one test will transfer nicely to another. Here, we will focus our example on the Bayesian *t*-test to test for a difference in means between two groups (i.e., an independent samples *t*-test), though the concepts, steps, and interpretations are essentially the same for other types of *t*-tests as well (e.g., single sample and paired-samples *t*-tests). We will get started quickly and exploit the similarity of the mechanics of using JASP to jump right into the test. But we will then need to spend a bit of time talking about how the Bayesian *t*-test works "under the hood". As in Chapter 3, I will also provide a detailed template that you can use for communicating the results of your own Bayesian *t*-tests.

EXAMPLE: PERCEPTION OF STEREOGRAMS

Have you ever seen something like Figure 4.1? This is called a *stereogram*, and it is a seemingly random image that actually hides a special treat. Specifically, it is constructed by masking a shape (in this case a knotted rope) with a repeated pattern in such a way that the target shape is only visible when viewed from just the right distance so that the eyes are focused on a point *behind* the visual plane of the stereogram (i.e., the page in the book). When this happens, the shape will magically appear to be floating in front of the page, producing a sensation of depth called stereopsis.

DOI: 10.4324/9781003470199-4

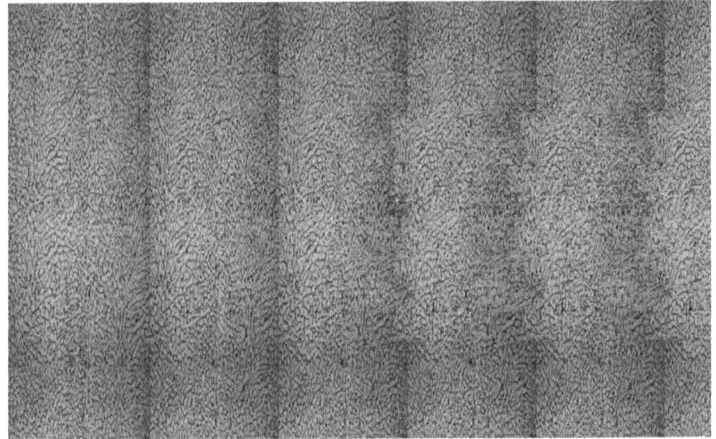

Figure 4.1 An example stereogram, constructed on https://easystereogrambu
ilder.com.

Try a while and see if you can see the knotted rope. If you can't,
don't worry. I've always been quite terrible with these things. There
are various other ways to help you perceive a stereogram, and a
quick online search can direct you to them. Fortunately, you do not
need to be able to "see" a stereogram to be able to do a Bayesian
t-test.

Instead, we are going to walk through an example of an experiment about stereograms. One classic example comes from a paper
by *Frisby and Clatworthy (1975)*, who presented random dot stereograms to 78 participants, each of whom was asked to indicate how
long it took them to see (or "fuse") the object. Thirty-five of the
participants were given extra visual information about the target
image (e.g., "The object-in-depth which you are to look for is a
spiral coming out of the screen towards you"; *Frisby & Clatworthy,
1975*, p. 175). The remaining 43 participants were given no help
and simply asked to press a button as soon as they saw the object. As
you might guess, Frisby and Clatworthy were interested in whether
giving participants extra visual information had an effect on the
fusion times.

The data from Frisby's and Clatworthy's study are available in the
JASP Data Library and may be accessed by navigating to section "2.

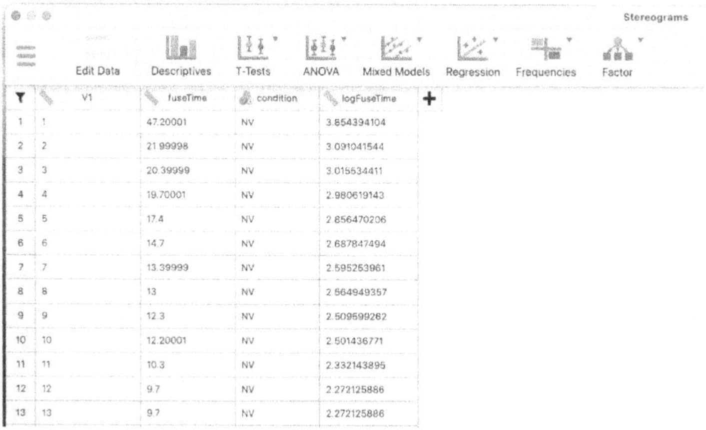

	V1	fuseTime	condition	logFuseTime	
1	1	47.20001	NV	3.854394104	
2	2	21.99998	NV	3.091041544	
3	3	20.39999	NV	3.015534411	
4	4	19.70001	NV	2.980619143	
5	5	17.4	NV	2.856470206	
6	6	14.7	NV	2.687847494	
7	7	13.39999	NV	2.595253961	
8	8	13	NV	2.564949357	
9	9	12.3	NV	2.509599262	
10	10	12.20001	NV	2.501436771	
11	11	10.3	NV	2.332143895	
12	12	9.7	NV	2.272125886	
13	13	9.7	NV	2.272125886	

Figure 4.2 A screenshot of the JASP data screen for the Stereograms dataset.

T-tests" and selecting the dataset "Stereograms". As we have done previously, you'll want to choose the option without the JASP logo. The dataset should appear similar to the one in Figure 4.2. Here, we have four variables: V1, which represents participant number; fuseTime, which represents the time (in seconds) that it took each participant to indicate that they saw the object; condition, which codes whether each participant was provided extra visual information (VV) or not (NV); and logFuseTime, which is a calculated variable giving the (natural) logarithm of the observed fuseTime measurements (note – the need for this specific variable will become apparent in a few moments).

Before jumping into an analysis of the data (in this case, an independent samples *t*-test), we should first explore our data and make sure that it meets the assumptions of the *t*-test. Recall that a *t*-test is used to compare the means of two samples that are each assumed to be drawn from a normal distribution. Let's see if our samples look like they meet this requirement. Click the "Descriptives" button and move fuseTime to the "Variables" box. Further, we can split the descriptives by condition by moving condition to the box labeled "Split". Additionally, you'll want to open the "Basic plots" menu and select "Distribution plots". Figure 4.3 shows the setup of our analysis options.

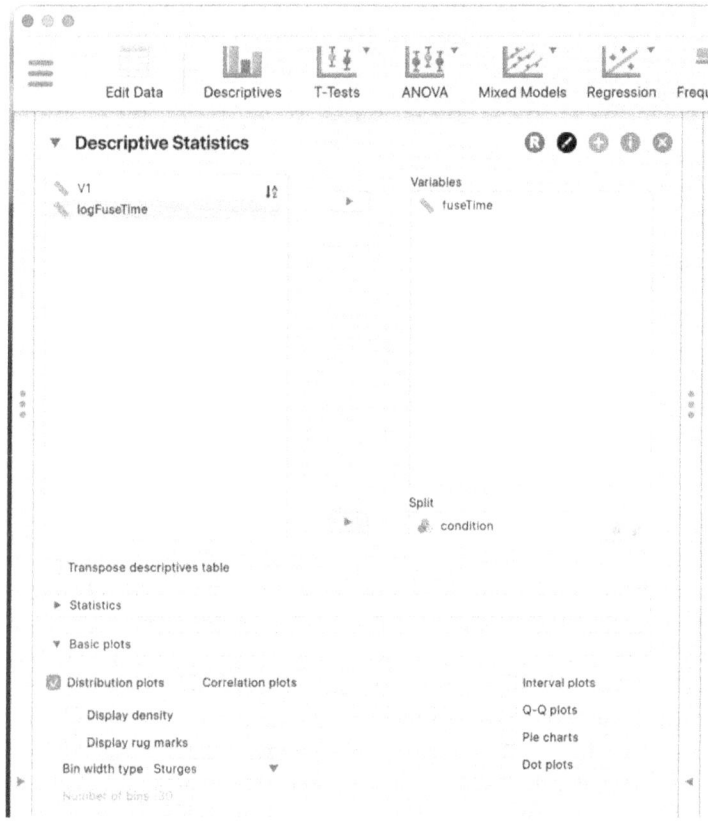

Figure 4.3 A screenshot of the JASP analysis screen for the Descriptives module, applied to the Stereograms dataset.

Immediately, you will see the descriptives output shown in Figure 4.4. Take a look at the distribution plots – they both appear to have a very pronounced rightward (positive) skew. You can confirm this visual intuition by additionally calculating the skewness of each sample. To do this, open the "Statistics" menu and select "Skewness" (this can be found in the list titled "Distribution"). The descriptives table in Figure 4.4 will then update to also show these skewness measures: for the NV group, the skewness is 2.873, and for the VV group, the skewness is 1.548. Certainly, this positive skew

Descriptive Statistics

Descriptive Statistics		
	fuseTime	
	NV	VV
Valid	43	35
Missing	0	0
Mean	8.560	5.551
Std. Deviation	8.085	4.802
Minimum	1.700	1.000
Maximum	47.200	19.700

Distribution Plots

fuseTime

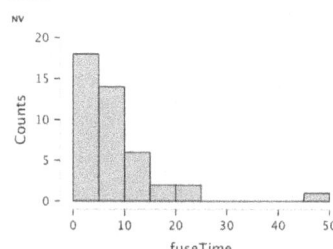

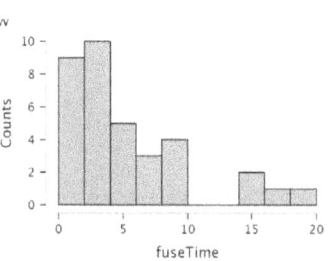

Figure 4.4 A screenshot of the JASP analysis screen showing distribution plots (i.e., histograms) for `fuseTime`.

indicates that the underlying populations from which these distributions were sampled are *not* normal, thus violating a key assumption of the *t*-test.

In cases like this, one option is to mathematically *transform* the offending measurements in such a way that will remove the skew. One way to do this is to take the *natural logarithm* of each measurement. If you've not encountered logarithms before, simply think of them as a tool for compressing datasets that are spread out. The logarithm changes exponential growth to linear growth. In the case of these distributions, the effect is to "pull in" the right tails of the distribution, thus reducing the skew and giving us distributions of (transformed) measurements that better meet the assumptions of the *t*-test.

While you can perform this calculation directly in JASP, our dataset already contains these logarithmically transformed fusion times as the variable `logFuseTime`. If we add this variable to our "Variables" box, we can see the resulting distribution is largely absent of this skew (see Figure 4.5). We can also confirm this by looking at the skewness values of our `logFuseTime` variable: for the NV

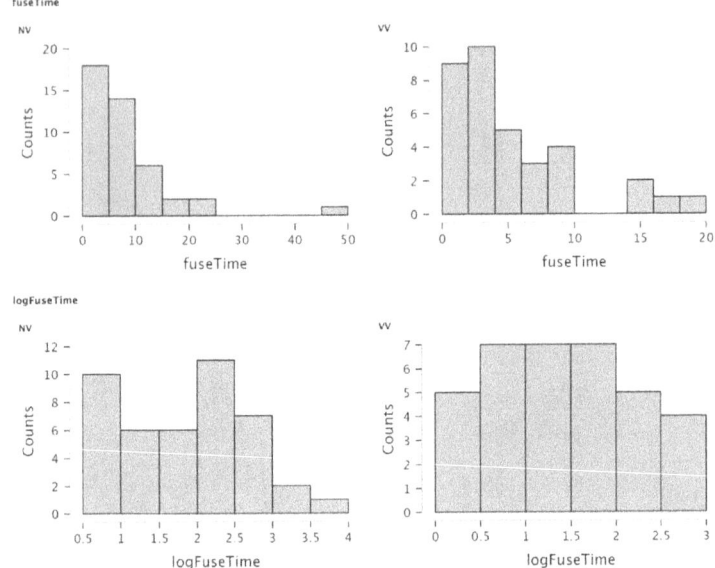

Figure 4.5 JASP screenshot comparing distribution plots for fuseTime and logFuseTime.

group, the skewness is 0.156, and for the VV group, the skewness is 0.189. Both of these are much closer to 0, which is the skewness value we would expect for a normal distribution. Thus, we will need to perform our *t*-test on the logFuseTime variable, not the original fuseTime measurements.

PERFORMING THE TEST IN JASP

Given our previous experience with performing tests in JASP, the most sensible thing to do at this point is to jump right in and try our hand at performing the Bayesian *t*-test. To that end, click the "T-Tests" button in the top menu bar and select the Bayesian independent samples *t*-test. As we discussed in the previous section, we'll be working with the log-transformed fuse times, so let's move the variable logFuseTime into the "Dependent Variables" box and move the variable condition into the "Grouping Variable" box. Also, let's go ahead and display the plots we've worked with

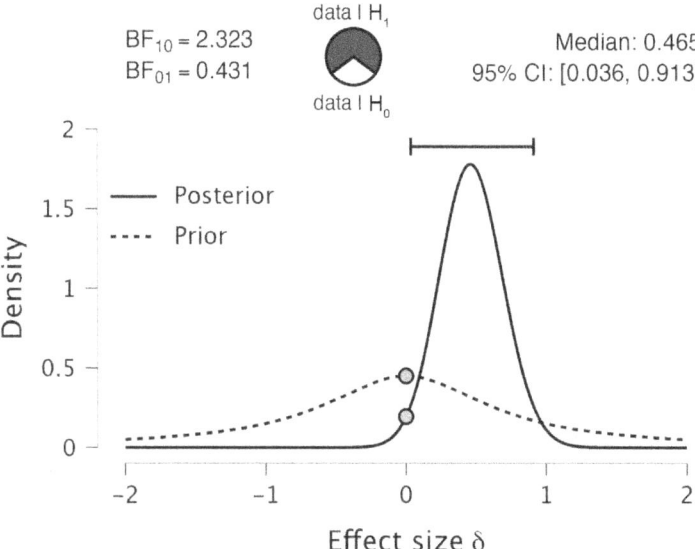

Figure 4.6 Prior and posterior plot for the Bayesian independent sample *t*-test.

before – namely, "Prior and posterior" and "Bayes factor robustness check". The output is displayed in Figure 4.6.

JASP provides us with familiar output that we can quickly interpret. From our experience with Bayesian correlations in Chapter 3, we can confidently go forward with a minimum of specific background knowledge about the mechanics of the *t*-test. First, we see a Bayes factor of $BF_{10} = 2.323$, which implies that the observed data are 2.323 times more likely under the alternative hypothesis $\mathcal{H}_1$ than under the null hypothesis $\mathcal{H}_0$. According to *Jeffreys' (1961)* evidence categories, these data provide anecdotal evidence for the alternative hypothesis. Further, we can convert this Bayes factor into a posterior model probability. Assuming 1:1 prior odds for $\mathcal{H}_1$ and $\mathcal{H}_0$, the posterior odds is now 2.323:1 in favor of $\mathcal{H}_1$, which is equivalent to a posterior probability of

$$p\left(\mathcal{H}_1 \mid \text{data}\right) = \frac{BF_{10}}{1 + BF_{10}} = \frac{2.323}{1 + 2.323} = \frac{2.323}{3.323} = 0.699.$$

Let's look closely at the prior and posterior plot. Here, we can see that the prior and posterior distributions quantify our uncertainty about the *effect size* δ. This is the population version of Cohen's d, which is defined as the difference between group means divided by the standard deviation. This plot shows us that our belief that $\delta = 0$ (i.e., the null hypothesis) decreases after observing data. Further, posterior distribution for δ has a median value of 0.465 (a "medium" effect according to *Cohen, 1988*), with a 95% credible interval of [0.036, 0.913]. Further, the sensitivity plot (see Figure 4.7) shows us that our qualitative assessment of "anecdotal" evidence persists across a wide range of prior settings.

In all, we can transfer most of what we learned about Bayesian tests from Chapter 3 into the context of this JASP output. However, we should now spend some time diving a bit deeper into the mechanics of the Bayesian t-test so that we can better appreciate what is happening under the hood.

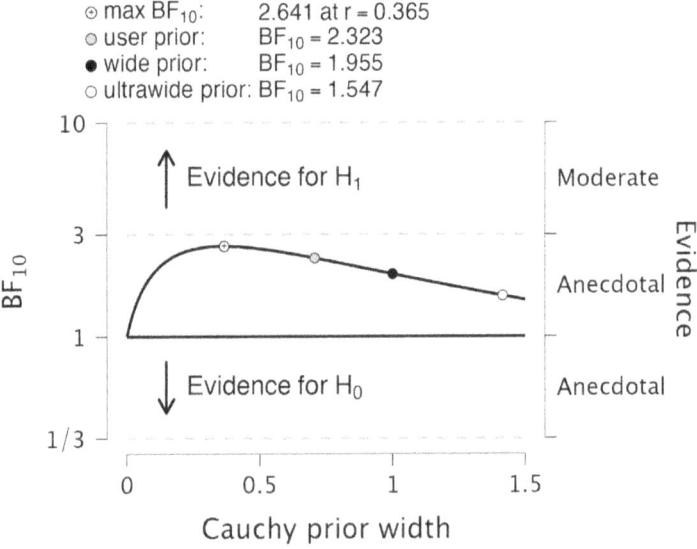

*Figure 4.*7 Sensitivity plot showing the Bayes factors obtained over a range of prior settings.

MECHANICS OF THE TEST

In this section, we will go into some detail and discuss the mechanics of the Bayesian *t*-test. During your first read of this book, it is perfectly fine to skip over this section, but I certainly recommend that you come back to it soon, because this is where we will answer some questions that you may already have about the JASP output that you saw in the previous example. For example, here are two questions that my own students usually have after performing the test:

1. What is that funny symbol δ, and what does it mean?
2. Why isn't the prior flat/uniform?

PARAMETERS FOR THE BAYESIAN *t*-TEST

First, let's address the symbol δ. As is common in inferential statistics, δ is a Greek letter corresponding to the "d" in the modern English alphabet and is pronounced "delta". As I indicated earlier in my explanation of our first example, δ represents the population-level *effect size*, or standardized difference between the two group means that we are comparing. If you know about effect size, especially in terms of Cohen's d, then you already have the right intuition. Mathematically, we can express the population effect size δ as

$$\delta = \frac{\mu_2 - \mu_1}{\sigma}$$

where μ_1 and μ_2 are the population means of groups 1 and 2, respectively, and σ is the population standard deviation of the groups. Recall that one of the assumptions of the *t*-test is that the groups have "common variance"; that is, their standard deviations are the same. Intuitively, δ represents the distance between the two group means μ_1 and μ_2 in terms of the common standard deviation σ. For example, $\delta = 1$ would mean that μ_2 is greater than μ_1 differ by exactly one standard deviation (i.e., σ). Symmetrically, $\delta = -1$ would mean that μ_2 is *less* than μ_1 by exactly one standard deviation. On the other hand, $\delta = 0$ would mean that the distance between μ_1 and μ_2 is 0 – that is, $\mu_1 = \mu_2$.

Beyond the fact that it makes good sense to use δ because of its relationship to Cohen's d, it is also convenient to use in our model definitions. As you may know, the null and alternative hypothesis in the classical t-test are usually defined as

- $\mathcal{H}_0 : \mu_1 = \mu_2$
- $\mathcal{H}_1 : \mu_1 \neq \mu_2$

for a two-sided test, and without loss of generality,

- $\mathcal{H}_0 : \mu_1 = \mu_2$
- $\mathcal{H}_1 : \mu_1 < \mu_2$

for a one-sided test. Using δ, we can rewrite these models in a quite useful way. For a two-sided test, the models would appear as

- $\mathcal{H}_0 : \delta = 0$
- $\mathcal{H}_1 : \delta \neq 0$

and for a one-sided test, they would be written as

- $\mathcal{H}_0 : \delta = 0$
- $\mathcal{H}_1 : \delta > 0$

This gives us a set of model definitions that fits nicely in the conceptual framework that we have learned from Chapters 2 and 3. First, the models we are comparing ($\mathcal{H}_0$ and $\mathcal{H}_1$) are expressed in terms of exactly one parameter δ. Second, the null hypothesis $\mathcal{H}_0$ is defined by restricting the parameter to a *single point* (in this case, $\delta = 0$). Third, the alternative hypothesis $\mathcal{H}_1$ is defined by letting the parameter vary over a range of values (for a two-sided test, we have $\delta \neq 0$, and for a one-sided test, we have $\delta > 0$ or $\delta < 0$).

Now that we know why JASP uses the funny symbol δ in the Bayesian t-test, we can address the second question. Why isn't the prior flat/uniform?

In Chapter 2, the binomial test used the parameter θ, which represents the probability of success on a single trial. As it is a probability, the value of θ can only live between the values of 0 and 1. In Chapter 3, the correlation test used the parameter ρ, which represents the Pearson correlation coefficient. Similarly, its value is also restricted – this time, between the values of -1 and $+1$.

For the Bayesian t-test, our parameter δ is *not* restricted. Theoretically, its values can take on *any* value, no matter how large. Of course, our experience as scientists will probably tell us that very large values of δ are not very probable, and it is much more likely that we will observe values of δ that are on the smaller side. Thus, when it is time to specify a prior distribution on δ, we need to find a way to mathematically encode this prior knowledge about δ. For the remainder of this section, we will discuss some ways to do this.

CONSTRUCTING A PRIOR FOR δ

Since the range of potential effect sizes δ is unrestricted, we will not be able to assign a uniform prior on δ. The reason for this is that a prior must be able to be defined as a probability distribution, and one of the properties of all probability distributions is that the area under the curve is equal to 1. For the binomial and correlation tests, we were able to use uniform distributions because the range of the test's parameter is bounded. For example, recall from Figure 2.2 that the uniform prior for the binomial test parameter θ has a height of 1 over the range of values extending from $\theta = 0$ to $\theta = 1$. The area under this curve (a rectangle) is equal to 1. Similarly, the uniform prior from Figure 3.1 for the correlation test parameter ρ has a height of 0.5 over the range of values from $\rho = -1$ to $\rho = 1$. The area under this curve (also a rectangle) is equal to $0.5 \times 2 = 1$.

However, we cannot make the prior for the t-test parameter δ a uniform distribution. No matter how hard we try, it is impossible to find a height h so that the area under the curve is equal to 1 over the range of possible values for δ. This is because the "width" of the range of values for δ is infinite. Thus, it is necessary that the height of the prior distribution will need to shrink to 0 as the values of δ become more extreme (i.e., as δ approaches infinity).

One way to accomplish this is to use what we already know. For example, what would happen if we just used a *standard normal distribution* for the prior on δ? As we can see in Figure 4.8, this prior is a standard bell-shaped curve centered at $\delta = 0$ with standard deviation $\sigma = 1$. Mathematically, this prior is permissible, because it is a probability distribution whose area under the curve is equal to 1.

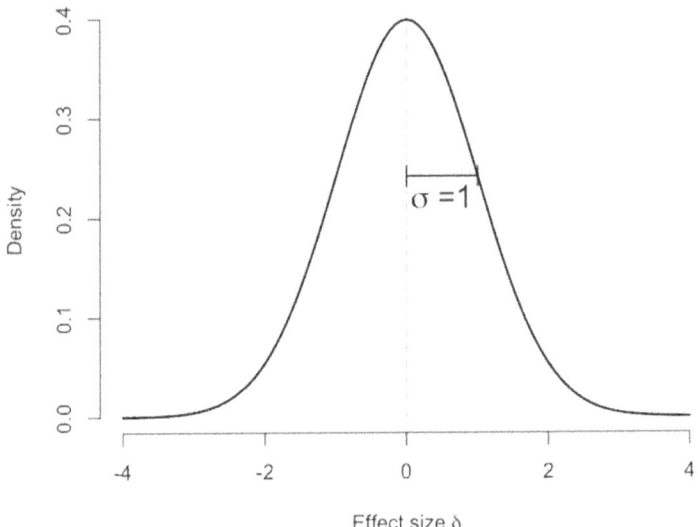

Figure 4.8 Placing a standard normal distribution as a prior on δ. This prior is often called the *unit information prior*.

Conceptually, this is also a useful prior for us because it nicely encodes our prior belief as scientists that most effect sizes should be small (i.e., ranging between -1 and $+1$), and the likelihood of encountering larger effect sizes decreases. Note that this prior is sometimes called the *unit information prior*, reflecting the non-obvious fact that, from an information-theoretic perspective, this prior contains the same amount of information as a single observation (*Kass & Wasserman, 1995*).

Though the unit information prior is a reasonable starting point for encoding our uncertainty about the effect sizes δ we expect to observe, it is slightly subjective in the sense that we chose the specific value 1 for the variance of the prior. One school of thought in Bayesian analysis is to construct *objective* priors that rely less on specific choices that we make in our analysis and work across a wider range of scenarios. One objective approach to constructing this normal prior for δ is to still center it at 0, but instead of fixing a value for the variance σ_{δ}^{2}, we allow this variance to vary. This additional level of uncertainty in σ_{δ}^{2} means that we must also place a prior on the variance σ_δ^2, thus giving our prior a *hierarchical* structure. This was the approach used by *Zellner and Siow (1980)*, who recommended using an inverse chi-square distribution as the prior for σ_δ^2.

Though this additional hierarchical structure that we have placed on our prior for δ may seem to have made things more complicated, there is a *trick* that we can employ. In Bayesian statistics, whenever an expression is built from a quantity that is assumed to be variable (e.g., by placing a prior on it), one can use calculus to compute the "average" of the expression over the variable quantity (we saw this when we talked about marginal probability in Chapter 2). This technique of "integrating out" the variance σ_δ^2 was considered by *Liang, Paulo, Molina, Clyde, and Berger (2008)*, who subsequently worked out that the corresponding prior for δ was equivalent to a *Cauchy*[1] distribution (see Figure 4.9). Compared to a normal distribution, a Cauchy distribution has much heavier tails, even though the area under the curve is still equal to 1. In fact, this "heavy tails" property makes the Cauchy distribution quite pathological, so much so that the Cauchy distribution does not have a mean or variance.

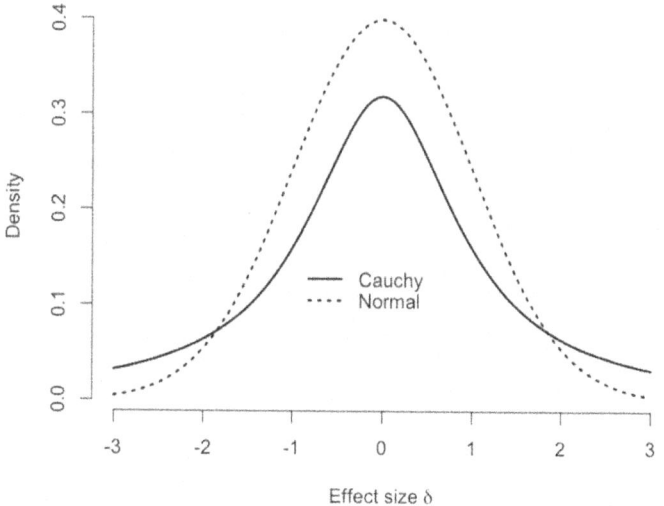

Figure 4.9 Compared to a normal distribution (the dashed line), a Cauchy distribution (solid line) has heavier tails, even though the area under the curve is still 1.

As is often the case in mathematics, seemingly new ideas are often really old ideas that have been rediscovered in a new context. The idea of placing a Cauchy prior on effect size δ was considered even earlier by Jeffreys in 1961. Because of the independent paths taken to constructing this objective, default prior for the t-test, the prior is often called the *JZS prior* (*Rouder, Speckman, Sun, Morey, & Iverson, 2009*), which acknowledges the multiple contributions to its construction, first of *Jeffreys (1961)*, and later *Zellner and Siow (1980)*.

In JASP, this JZS prior is exactly the one that is used by default. In fact, JASP uses a *scaled* Cauchy prior, where the setting of the scale r is equivalent to specifying the interquartile range of the distribution. By default, JASP uses the scale $r = 1/\sqrt{2} = 0.707$ – this means that 50% of the expected effect sizes would be less than 0.707 in magnitude. While r can be varied continuously and given any value $r > 0$, two common settings used by JASP in its sensitivity analyses are $r = 1$ (a "wide" prior) and $r = \sqrt{2} = 1.414$ (an "ultra-wide" prior). You can see these values of r and the corresponding labels in the Bayes factor robustness plot that we constructed in our earlier example. Additionally, Figure 4.10 shows these priors all

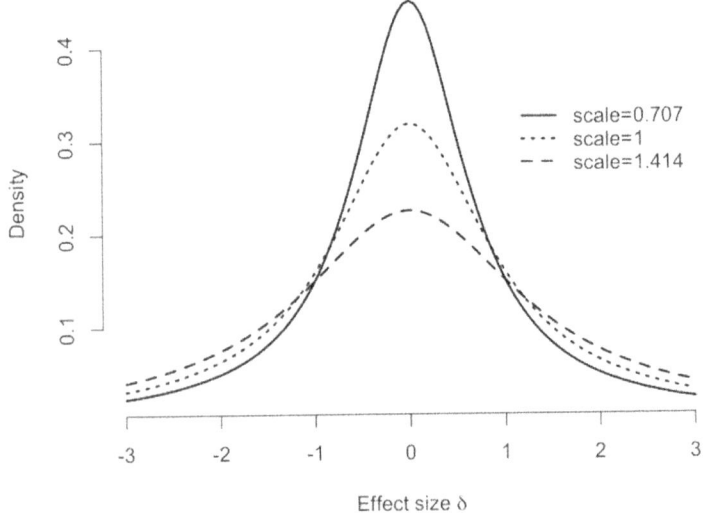

Figure 4.10 Three Cauchy distributions of differing scale, each corresponding to typical prior settings in JASP.

in one plot so that you can see how varying the scale changes the overall shape of the Cauchy prior.

How to communicate the results

As we did in Chapter 3, we'll finish up our example with a discussion of how to communicate the results of this Bayesian analysis. Just like with the Bayesian correlation, there are three main components that should be included:

1. Model definitions
2. Model comparison and parameter estimation
3. Sensitivity analysis.

Model definitions: The first step of reporting any Bayesian analysis is to include a definition of the models being considered. Importantly, we must additionally specify the prior distribution under $\mathcal{H}_1$. Here is an example of what this could look like for the *t*-test we just performed:

> *The parameter of interest in the Bayesian independent samples t-test is the population effect size δ. Under the null hypothesis, we expect an effect size of $\delta = 0$ and thus define $\mathcal{H}_0 : \delta = 0$. Under the alternative hypothesis, the effect size δ takes on non-zero values; i.e., $\mathcal{H}_1 : \delta \neq 0$. We specify our uncertainty about δ by defining the prior distribution for δ under $\mathcal{H}_1$ to be a Cauchy distribution with scale $r = 0.707$, which is the default prior used in JASP.*

Model comparison and parameter estimation: Next, we should report the results of our model comparison and parameter estimation. Recall from our discussion in Chapter 3 that we should perform the model comparison *first*, and then if warranted, proceed with the parameter estimation from the posterior distribution. Here is an example from our analysis:

> *We found a Bayes factor of $\mathrm{BF}_{10} = 2.323$, which means that the observed data are approximately 2.3 times more likely under $\mathcal{H}_1$ than under $\mathcal{H}_0$. Using the suggested evidence categories from Jeffreys (1961), this result indicates anecdotal evidence in favor of $\mathcal{H}_1$. The posterior*

distribution for δ had a median of $\delta = 0.465$, with a 95% credible interval that ranges from 0.036 to 0.913.

Further, as the Bayes factor represents the updating factor on prior model odds, we can translate it directly into a posterior model probability. Indeed, if we begin with 1:1 prior odds for $\mathcal{H}_1$ and $\mathcal{H}_0$ (i.e., assuming $\mathcal{H}_1$ and $\mathcal{H}_0$ are equally likely, a priori), the posterior odds are 2.323:1 in favor of $\mathcal{H}_1$. This is equivalent to a posterior model probability[2] of $p(\mathcal{H}_1 \mid \text{data}) = 0.699$.

Sensitivity analysis: Finally, we need to describe how our inference depends on our specific choice of prior distribution for the parameter δ. An example of this might look like the following:

Since Bayes factors depend on the specific choice of prior distribution for our model parameter δ, we additionally performed a sensitivity analysis to assess the impact of this prior choice on our inference. In Figure 4.7, we have plotted the Bayes factor as a function of the Cauchy prior scale r, which is the parameter that governs the width of our prior distribution. We can see that across a wide range of reasonable values for this scale parameter, the obtained Bayes factor represents anecdotal evidence in favor of $\mathcal{H}_1$.

Directional tests

Our original question was whether giving people extra visual information had an effect on the fuse times of stereograms. As stated, this is a *nondirectional* test; that is, there is no predicted direction for the effect. No matter whether we observe an increase or a decrease in the fuse times, both results would be characterized as "an effect". However, we might instead be interested in whether the presentation of extra visual information indeed *decreases* fuse times. In this case, we would want to perform a *directional* test, also known as a *one-tailed test*. In this section, we'll discuss how to perform such a test in JASP, and then we'll discuss how the results we obtain are related to the original nondirectional test that we performed previously.

As one might expect, it is quite easy to perform a directional test in JASP. The key difference is in the specification in the list under "Alternative hypothesis" – this time, instead of the default option

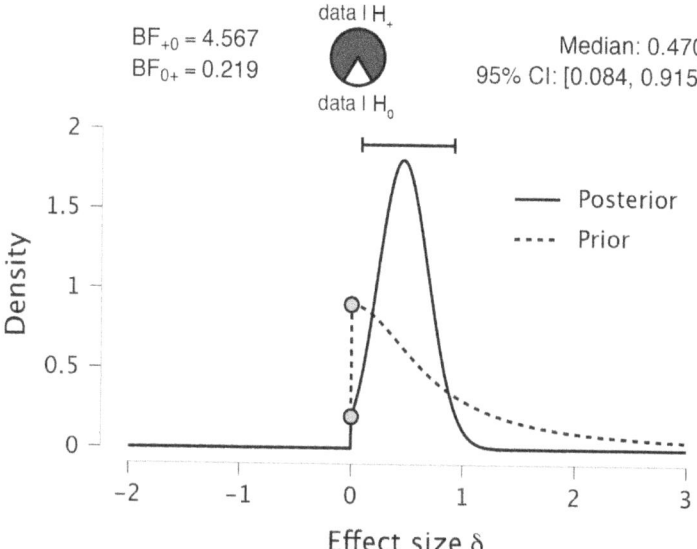

Figure 4.11 Prior and posterior plot for the directional Bayesian *t*-test.

"Group 1 ≠ Group 2", you'll want to select the second option "Group 1 > Group 2". Doing so will immediately update the output in JASP in several ways. First, you'll notice that the notation of the Bayes factor changes from BF_{10} to BF_{+0}, and moreover, the value of this Bayes factor increases to $BF_{+0} = 4.567$. The resulting prior and posterior plot and sensitivity analysis are depicted in Figures 4.11 and 4.12. This leads to two immediate questions: (1) what is this new notation BF_{+0}; and (2) why did the Bayes factor get larger?

First, let's address the change in notation. BF_{+0} is simply a notation to represent the Bayes factor for a directional hypothesis. The "+" in this case represents the model $\mathcal{H}_+ : \delta > 0$, where the difference is taken as the mean of Group 1 (the NV group) minus the mean of Group 2 (the VV group). A quick side note: if you're ever unsure of which group JASP is considering as Group 1 and Group 2, just click "Descriptives" under "Additional Statistics". The resulting table will list the groups in the order assumed by JASP, so that the first row represents Group 1, the second row represents Group 2, etc.

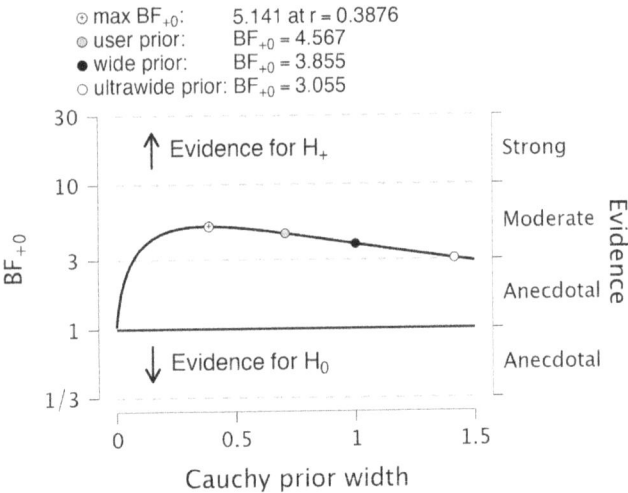

Figure 4.12 Sensitivity plot for the directional Bayesian *t*-test, showing the Bayes factors obtained over a range of prior settings.

Second, let's address why the observed Bayes factor for the directional test is larger than the one we observed for the nondirectional test. I'll try to explain this in two different ways. The first of these will be an explanation based on intuition that we can gain from the prior and posterior plot, and the second will be based on some more formal mathematical reasons.

Let us consider the prior and posterior plot that we just obtained. You'll notice immediately that the prior distribution looks different from the prior used in the original nondirectional test. This is because when we perform a directional test (in this case, imposing the inequality constraint $\delta > 0$), we are inherently specifying a "truncated" prior distribution to quantify our a priori uncertainty about δ. But we cannot just remove the left half of the distribution, for doing so would leave a distribution whose area under the curve is half of the original. Because probability distributions must have a total area under the curve of 1, we must take all of the mass that was contained in the left side of the prior and somehow put it into the right side. A way to think of this is "folding over" the prior onto itself. Doing so will then double the height of the prior distribution at each point δ. This can be easily seen at $\delta = 0$, where the height in the original prior was approximately 0.5, and the height in this new "folded over" or truncated prior is approximately 1.

Now, remember that the Bayes factor can be visualized on this plot as the ratio of the heights of $\delta = 0$ in the prior and posterior (i.e., using the Savage-Dickey density ratio). As we just discussed, the height of $\delta = 0$ in the prior has doubled. But as we can confirm by looking at the median and 95% credible interval of the posterior, the posterior distribution hasn't changed all that much. Thus, the height of $\delta = 0$ in the posterior also hasn't changed that much. In total, then, the change in the value of the Bayes factor is governed primarily by the change in the height of $\delta = 0$ in the prior – i.e., it has (approximately) doubled.

If you are not completely satisfied with this intuitive explanation, consider the following more formal mathematical justification. To compute the directional Bayes factor BF_{+0}, we can use the arithmetic property of *transitivity* as follows. First, express BF_{+0} according to its definition:

$$\text{BF}_{+0} = \frac{p\left(\text{data} \mid \mathcal{H}_+\right)}{p\left(\text{data} \mid \mathcal{H}_0\right)}.$$

Now we'll perform a little trick – if we multiply by something that is equal to 1, we don't change the equality of the statement. In this case, I'm going to multiply by a fraction whose numerator and denominator are equal:

$$\text{BF}_{+0} = \frac{p\left(\text{data} \mid \mathcal{H}_+\right)}{p\left(\text{data} \mid \mathcal{H}_0\right)} \cdot \frac{p\left(\text{data} \mid \mathcal{H}_1\right)}{p\left(\text{data} \mid \mathcal{H}_1\right)}.$$

Let's rearrange some terms of the right-hand side:

$$\text{BF}_{+0} = \frac{p\left(\text{data} \mid \mathcal{H}_+\right)}{p\left(\text{data} \mid \mathcal{H}_1\right)} \cdot \frac{p\left(\text{data} \mid \mathcal{H}_1\right)}{p\left(\text{data} \mid \mathcal{H}_0\right)},$$

which we can rewrite in "BF" notation as follows:

$$\text{BF}_{+0} = \text{BF}_{+1} \cdot \text{BF}_{10}.$$

Thus, we can see that the directional Bayes factor BF_{+0} can be obtained by multiplying the nondirectional Bayes factor BF_{10} by some correction factor, which is BF_{+1}. This "correction" Bayes factor is not trivial to compute, but *Klugkist, Kato, and Hoijtink (2005)* showed that it can be computed as the proportion of the posterior distribution, which satisfies the inequality constraint $\delta > 0$ divided by the proportion of the prior distribution which satisfies the same inequality constraint. As we can see in our original prior and posterior plot (i.e., the one from the nondirectional test), almost all the posterior distribution satisfies $\delta > 0$. But, as the Cauchy prior is symmetric around $\delta = 0$, exactly half of the prior satisfies $\delta > 0$. Thus, we can see that the correction factor is approximately 1 divided by 1/2, which is approximately 2. That is, the directional Bayes factor BF_{+0} is approximately 2 times the nondirectional Bayes factor BF_{10}.

CHAPTER SUMMARY

Let's recap what we've learned in this chapter:

1. We introduced the Bayesian independent samples t-test, which compares two models about the population effect size δ: a null hypothesis $\mathcal{H}_0$, where $\delta = 0$, and an alternative hypothesis $\mathcal{H}_1$, where δ is unrestricted and varies according to a prior distribution on δ.

2. We discussed the development of the *Cauchy distribution* as a default prior for the population effect size δ. This prior has one *scale* parameter r, which is called the "width" of the prior. This parameter has a default value of $r = 0.707$, and can be thought of as defining the interquartile range of the Cauchy prior. That is, $r = 0.707$ means that 50% of the a priori observed effect sizes should be between -0.707 and 0.707.

3. We worked through a detailed example in JASP and gave an example of how to communicate the results of our Bayesian t-test.

4. We concluded with a discussion of how to construct one-sided, directional versions of the Bayesian t-test. Intuitively, the test works by folding over one side of the Cauchy prior onto the other.

EXERCISES

1. This exercise will use the "Directed Reading Activities" dataset from the JASP Data Library. A teacher believes that directed reading activities in the classroom can improve the reading ability of elementary school children. She convinces her colleagues to give her the chance to try out the new method on a random sample of 21 third-graders. After they participated for eight weeks in the program, the children take the Degree of Reading Power test (DRP). Their scores are compared to a control group of 23 children who took the test on the same day and followed the same curriculum apart from the reading activities.

   a. Create a table that shows the means and standard deviations of DRP scores in the control and treatment group.

   b. Play around with the plots that JASP offers in the Descriptives module to produce a nice boxplot showing the distribution of DRP scores for each group.

   c. Perform a Bayesian independent samples t-test to test whether the control group performs *worse* than the treatment group. How do you interpret the Bayes factor you obtained in this analysis?

   d. Perform a sensitivity analysis. To what extent does your inference depend on the specific choice of prior scale?

2. *Borota et al. (2014)* observed that in a sample of 73 participants, the 35 participants who received 200 mg of caffeine had significantly better scores on a test for memory of objects than did the 38 participants who took a placebo, $t(71) = 2.0$, $p = 0.049$. *Borota et al. (2014)* concluded that caffeine enhances memory consolidation.

   a. Using the JASP Summary Statistics module, perform a Bayesian reanalysis of this claim. What do you find?

   b. Note that the Bayes factor depends on the choice of prior on the population effect size δ. Are there any choices for prior that would result in the data becoming evidential for the null?

   c. Given these results, how confident are you of Borota et al.'s claims?

NOTES

1 Named after the French mathematical Augustin-Louis Cauchy (1789-1857), usually pronounced "co-SHEE".

2 Recall that $p\left(\mathcal{H}_1 \mid \text{data}\right) = \dfrac{BF_{10}}{1 + BF_{10}} = \dfrac{2.323}{1 + 2.323} = \dfrac{2.323}{3.323} = 0.699$.

BAYESIAN ANALYSIS
OF VARIANCE

And now for something completely different!

After reading the first four chapters of this book, I am hopeful that you have gotten some intuition about how Bayesian inference works. The Bayesian versions of the correlation test and the t-test are reasonably straightforward to understand and easy to perform in JASP. They are also good subjects with which to begin a study of Bayesian statistics because their prior distributions are easy to describe, and the JASP output gives a good visualization of the entire Bayesian updating process. From here, our next target is the analysis of variance, or ANOVA for short.

In many of his papers, mathematical psychologist and Bayesian statistician Jeff Rouder refers to the analysis of variance as the "workhorse" of experimental designs (e.g., *Rouder, Engelhardt, McCabe, & Morey, 2016*). Indeed, when students take courses in experimental design, variations on the classical analysis of the variance are the most often used methods throughout the course. During my own graduate training, some of the courses in applied statistics were simply referred to by the colloquial phrase "the ANOVA course". Simply put, analysis of variance is a fundamental tool for analyzing data from experiments. Thus, it is worthwhile that we spend some time in this chapter learning about a Bayesian approach to performing analysis of variance.

While I would like to spend most of my time describing Bayesian analysis of variance, for background we should briefly discuss the classical approach. Though some readers will have experience with classical ANOVA, even the inexperienced reader will benefit from a conceptual description of how this classical approach works.

DOI: 10.4324/9781003470199-5

Though the Bayesian approach to ANOVA is fundamentally based on model comparison, and thus differs from the classical approach, there are Bayesian tools for reporting these model comparisons that mirror the classical decomposition into "main effects" and "interactions" that pervades the scientific literature. Thus, it is important that we review these terms before diving into our own example of performing a Bayesian ANOVA.

A CLASSIC ANOVA EXAMPLE

For a working example, we will use the classic "Tooth Growth" data set that is included by default in most statistical software packages, including JASP. These data were originally published in *Bliss (1952)*, who used data obtained by *Crampton (1947)* in an experiment on the effect of vitamin C supplementation on tooth development in guinea pigs. In his study, Crampton gave 60 guinea pigs a dietary supplement in one of two forms: orange juice, or an ascorbic acid supplement. Further, the dosage of the vitamin C supplement was varied; the guinea pigs were either given 500, 1000, or 2000 micrograms per day. After about a month and a half on the diet, tooth growth was measured by removing the incisors and sectioning them to measure the length (in microns) of the odontoblasts, a cell responsible for tooth growth.

We can load this dataset into JASP in the usual way. Under the JASP Data Library, select category "3: ANOVA" and find the dataset titled "Tooth Growth". As in our past chapters we want the one *without* the JASP logo. The resulting data set should have 3 columns:

- len, a continuous variable representing the observed length of each guinea pig's odontoblast, measured in microns.
- supp, a categorical variable representing the type of vitamin C supplement administered. Those given orange juice are labeled OJ, whereas those given the ascorbic acid supplement are labeled VC.
- dose, a categorical variable representing the amount of supplement given to each guinea pig per day – either *500, 1000*, or *2000* micrograms.

As an experiment, these data represent a classic *2 x 3 factorial design*. That is, there are two independent variables (supp and dose), where supp has two levels and dose has three levels. Both independent variables (or *factors*) are manipulated *between subjects*, which means that each of the 60 guinea pigs receives exactly one combination of supp and dose. Moreover, it is a *balanced* design, meaning that each factor is manipulated so that all experimental conditions have the same number of observations. That is, for the supp factor, exactly half ($N = 30$) received OJ, whereas the other half ($N = 30$) received VC. Similarly, for the dose factor, exactly one-third ($N = 20$) receive 500, 1000, and 2000 micrograms of supplement, respectively.

Before diving into a Bayesian analysis of variance with these data, let's briefly describe the classical ANOVA approach and show how to perform it in JASP. At its core, classical analysis of variance is a technique that works by decomposing the total observed variance in a dataset into multiple sources – the variance due to some experimental manipulation, and the leftover, or *residual*, variance. In the case of a factorial design, there are multiple possibilities for the types of experimental effects we might observe. As shown in Figure 5.1, the first stage of this variance decomposition is to split the total variance into two sources:

1. The variance between the experimental treatment groups, which is roughly equal to the variance of the group means from the overall grand mean
2. The leftover, residual variance (i.e., everything left after you subtract out the variance between treatments.

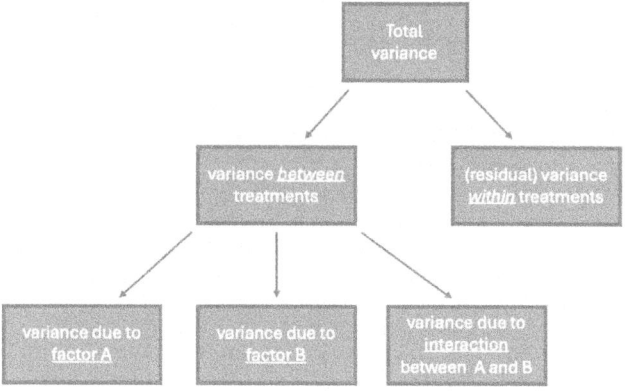

Figure 5.1 A diagram showing the decomposition of total variance in classical ANOVA.

Next, we split the variance between treatments into three possible sources: (1) the variance due to the manipulation of the first factor (labeled factor A), (2) the variance due to the second factor (labeled factor B), and (3) the variance due to the interaction between factors A and B. If there is meaningful variance due to either factor A or B, these effects are called *main effects*, and they represent a difference in the means observed after collapsing across the levels of the other variable. On the other hand, an *interaction effect* represents a difference in the pattern of means in one variable that depends on the level of the other variable.

Let's make this a bit more concrete. In JASP, look for the "ANOVA" button in the top menu bar. Click it and select "ANOVA" under the list of "Classical" analyses. In the analysis screen, move `len` into the "Dependent Variable" box and move both `supp` and `dose` into the "Fixed Factors" box. The ANOVA table will quickly appear in the results screen and is shown in Figure 5.2.

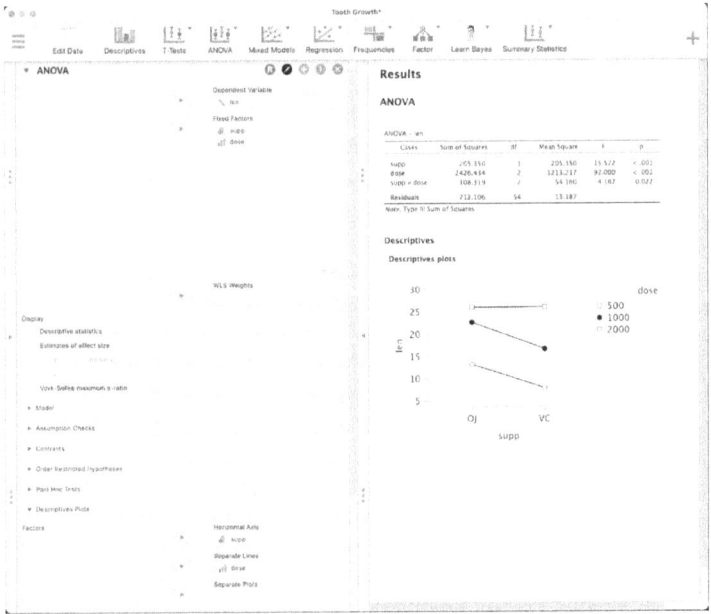

Figure 5.2 A screenshot of the classical ANOVA procedure in JASP.

Under the column labeled "Cases", you'll see the sources of variability that we just described: (1) the main effect of supp; (2) the main effect of dose; (3) the interaction between supp and dose, labeled supp*dose; and (4) the residual variance. Statistically, the ANOVA works by using as a test statistic the ratio of the decomposed sources of treatment variance to the leftover, residual variance. Under the null hypothesis, this test statistic is distributed as an F-distribution, and the resulting p-value represents the likelihood of observing the given F-ratio or more extreme under the null hypothesis. In this analysis, all three p-values are small. Following our typical logic for p-values, we reject $\mathcal{H}_0$ for all three, giving us "significant" main effects of supp and dose, as well as a significant interaction.

What does all this mean? Perhaps the best way to see that is to produce a plot of the condition means, which will clearly show each of these effects. To produce this plot in JASP, find the menu for "Descriptives Plots" in the analysis pane. To produce an effective plot of condition means for a factorial design, we can plot one of the factors on the horizontal axis, and for the other factor, we can have a different line for each of the factor's levels. In Figure 5.2, you can see that I have chosen to represent supp on the horizontal axis and to represent the levels of dose as separate lines.

The resulting plot will immediately appear on the results screen (see Figure 5.2). Let's look at how it represents each of the main effects and the interaction. First, let's consider the main effect of dose. Notice how the three levels of dose are represented as three separate lines in the plot? You may also notice also that these lines have a lot of vertical separation. Specifically, the mean odontoblast lengths for guinea pigs (ignoring the specific value of supplement type supp) are largest for guinea pigs given 2000 micrograms per day, followed by those given 1000 and 500 micrograms, respectively. This observed vertical separation is a clear visualization of the main effect of dose.

Now let's tackle the main effect of supp. This one is a little harder to see. Here, we are interested in the mean odontoblast length for the guinea pigs given OJ versus those given VC, ignoring the specific dosage given. Visually, we need to "collapse" (i.e., take the mean of) the three different doses that we see stacked above OJ, as well as the three we see above VC. When we do this, we see that the mean length is larger for OJ than for VC – this difference in mean lengths is exactly what we mean by "*the main effect of* supp". In both cases, the main effect can be thought of as a difference in means for one factor when we collapse across the levels of the other factor.

The interaction is also quite easy to see. The idea of an interaction is that the pattern of observed means in one variable will differ when you change the level of the other variable. In this case, consider the pattern of means we observe when comparing lengths due to supp (i.e., OJ versus VC). For two of the doses (500 and 1000), we see that the mean length for OJ is larger than the mean length for VC. However, for the third dose (2000), there is no difference in mean lengths between OJ and VC. That is, the pattern of observed means in supp differs across the levels of dose. This is what we mean by an interaction.

BAYESIAN ANALYSIS OF VARIANCE

The first time I tried a Bayesian analysis of variance myself, I was struck at how unfamiliar it was. This is because, at first, the Bayesian analysis of variance looks completely different from the classical version of ANOVA that we reviewed earlier. Like our previous examples, the Bayesian ANOVA works by comparing the predictive performance of different competing models (*Rouder et al., 2016*). But, in contrast to the relatively simple procedures of Chapters 3 and 4 where there are only two models to compare ($\mathcal{H}_0$ versus $\mathcal{H}_1$), the Bayesian analysis of variance simultaneously considers the performance of *several* models.

To explain, let's consider Figure 5.3. Here, we have *five* models, each representing one of the unique combinations that can be made with two independent variables in a factorial design. First, we have the *null model*, which predicts all observed data with the grand mean alone. That is, there are no effects of any of the manipulated independent variables. Next, we have the *single main effect models*. These models predict the observed data according to their group membership for *either* factor A or B, but not both. If *both* factor A and B have main effects on the dependent variable, this situation is represented by the *additive model*. Finally, if we have both main effects *and* an interaction, the *interactive* model will be the correct one.

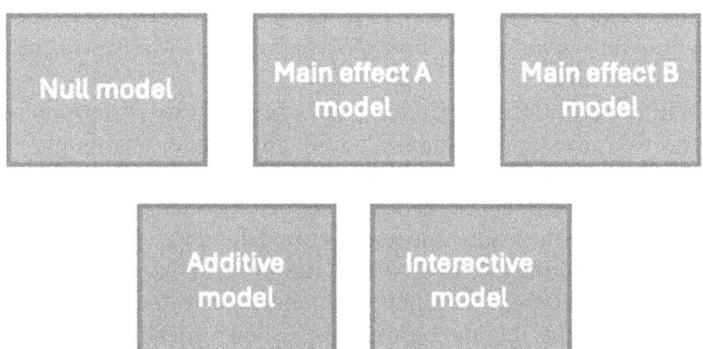

Figure 5.3 A schematic representation of the five possible models arising from a 2x2 factorial design.

You may be wondering why some combinations of main effects and/or interactions are missing. For example, what about a model that proposes a main effect of factor A and an interaction, but no main effect of factor B? The answer is complicated. While this model (and others like it) can certainly be proposed, there are some principled reasons for not including them in our model list. The primary reason has to do with something called the *principle of marginality*. The idea is that any linear model that contains an interaction should contain all terms that are contained in the interaction. Thus, the model just described would not be included our model list because it doesn't contain a term for the main effect of factor B. While some have argued for enforcing the marginality principle in Bayesian ANOVA (e.g., *Rouder et al., 2016*; *Nelder, 2000*), others have argued to relax it (e.g., *Heathcote & Matzke, 2021*). Nonetheless, the default behavior in JASP is to enforce marginality, so I will follow that path here.

Now that we've established the general approach of the Bayesian analysis of variance, let's discuss how it works in JASP. In previous chapters, I would simply recommend to just perform the test – after all, JASP's Bayesian tests are just as easy to perform as the classical tests. However, doing that in the case of the Bayesian ANOVA could be a step too far, as my earlier parable demonstrates. So to head off this potential obstacle (and prevent the reader from possibly turning away from Bayesian statistics completely!), I'd like to spend a little time talking about how JASP approaches the Bayesian analysis of variance.

In the previous Bayesian tests we performed (the correlation test and the independent samples t-test), the output was simple – a Bayes factor for $\mathcal{H}_1$ over $\mathcal{H}_0$. Certainly, in these cases, that is the primary output we sought, as there were only two models under contention, and thus only one possible model comparison that could take place. In the case of ANOVA, however, we have five different models. Moreover, the number of models increases exponentially as you increase the number of independent variables. So figuring out how to organize the *many* possible model comparisons that can occur here is something worth thinking about. The way the JASP software developers have done this is, I think, quite a nice way to summarize the output, so let's explore that here.

The structure of the JASP output can be seen in Table 5.1, which presents the output of a hypothetical Bayesian ANOVA. The idea is that each of the five models is first assigned a prior model probability, represented in the column labeled $p(\mathcal{M})$. For example, we could choose to assign equal probabilities to all five models, thus giving each model $\mathcal{M}$ a prior model probability of $p(\mathcal{M}) = 1 / 5 = 0.20$.

Then, JASP uses the BayesFactor package in R to compute Bayes factors for each model over the null model. This is represented by the column labeled BF_{10}. I've updated Table 5.2 to reflect these (again, hypothetical) Bayes factors.

Given the prior model probabilities $p(\mathcal{M})$ and the Bayes factors BF_{10}, the next step is to compute the *posterior model probabilities* $p(\mathcal{M} \mid \mathrm{data})$. This is easy, but tedious (so thankfully JASP does this for us automatically). The basic idea is to recall from Chapter 2 that

Table 5.1

Model	$p(\mathcal{M})$	$p(\mathcal{M} \mid \mathrm{data})$	$\mathrm{BF}_{\mathcal{M}}$	BF_{10}
Null	0.20			
A	0.20			
B	0.20			
A+B	0.20			
A+B+A$\star$B	0.20			

Table 5.2

Model	$p(\mathcal{M})$	$p(\mathcal{M}\,|\,\text{data})$	$\text{BF}_{\mathcal{M}}$	BF_{10}
Null	0.20			1
A	0.20			2
B	0.20			5
A+B	0.20			10
A+B+A⋆B	0.20			25

posterior *odds* are equal to the prior odds multiplied by the Bayes factor. Thus, for each model $\mathcal{M}_i$, we have

$$\frac{p(\mathcal{M}_i\,|\,\text{data})}{p(\mathcal{M}_0\,|\,\text{data})} = \frac{p(\mathcal{M}_i)}{p(\mathcal{M}_0)} \times \text{BF}_{i0}.$$

For example, the interactive model $\mathcal{M}_4$ has a Bayes factor of 25 over the null model. This would translate to the following equation:

$$\frac{p(\mathcal{M}_4\,|\,\text{data})}{p(\mathcal{M}_0\,|\,\text{data})} = \frac{0.20}{0.20} \times 25 = 25.$$

Of course, this equation has two unknowns (the posterior probabilities of the interactive model $\mathcal{M}_4$ and the null model $\mathcal{M}_0$), so we cannot find unique values for those two posterior probabilities. However, note that we will get a similar equation for each of our non-null models $\mathcal{M}_1,\ldots,\mathcal{M}_4$. Plus, we also know that the posterior probabilities must add to 1. This gives us a total of five equations with five unknowns (the posterior probabilities of each model). Doing some (albeit, very tedious) algebra will give us unique solutions for each posterior probability. I've gone ahead and found those posterior probabilities and included them in the updated Table 5.3.

Remember that Bayesian inference is all about how data updates our prior belief into posterior belief. In Table 5.3, this prior belief is represented by the column of prior model probabilities $p(\mathcal{M})$, and this posterior belief is represented by the posterior model probabilities $p(\mathcal{M}\,|\,\text{data})$. The first thing one might notice about these

Table 5.3

Model	$p(\mathcal{M})$	$p(\mathcal{M} \mid \text{data})$	$\text{BF}_{\mathcal{M}}$	BF_{10}
Null	0.20	0.023		1
A	0.20	0.047		2
B	0.20	0.116		5
A+B	0.20	0.233		10
A+B+A⋆B	0.20	0.581		25

posterior probabilities is that only two of them reflect an *increase* in belief after observing data – the additive model $A + B$ and the interactive model $A + B + A\star B$. The remainder reflect a *decrease*, indicating that observing data has, in fact, *diminished* our belief in the null model and the single factor models.

The remaining column in the table, $\text{BF}_{\mathcal{M}}$, serves as an index of this updating process for each model. The column $\text{BF}_{\mathcal{M}}$ represents the update in *model odds* from prior to posterior. That is,

$$\frac{p(\mathcal{M} \mid \text{data})}{1 - p(\mathcal{M} \mid \text{data})} = \frac{p(\mathcal{M})}{1 - p(\mathcal{M})} \times \text{BF}_{\mathcal{M}}.$$

At first, it may be difficult to see how this Bayes factor $\text{BF}_{\mathcal{M}}$ is different from the Bayes factor BF_{10}. The latter Bayes factor (BF_{10}) represents the likelihood of the observed data under the given model compared to the *null* model. As such, the Bayes factors in this column always represent a comparison to the null model $\mathcal{M}_0$. On the other hand, the Bayes factors in the column $\text{BF}_{\mathcal{M}}$ represent the data-driven *updating factor* for each model. Let's walk through the computation of one of these Bayes factors – say, the interactive model $A + B + A\star B$. By definition, $\text{BF}_{\mathcal{M}}$ is the factor by which the prior odds are multiplied to get posterior odds. Equivalently, it is what we get when we take the posterior odds for the interactive model and divide by the prior odds. So let's calculate these odds. The posterior odds can be computed as

$$\frac{p(\mathcal{M} \mid \text{data})}{1 - p(\mathcal{M} \mid \text{data})} = \frac{0.581}{1 - 0.581} = \frac{0.581}{0.419} = 1.387.$$

Similarly, the prior odds can be computed as

$$\frac{p(\mathcal{M})}{1-p(\mathcal{M})} \times \mathrm{BF}_{\mathcal{M}} = \frac{0.20}{1-0.20} = \frac{0.20}{0.80} = 0.250.$$

Taking their quotient, we get

$$\mathrm{BF}_{\mathcal{M}} = \frac{\text{posterior odds}}{\text{prior odds}} = \frac{1.387}{0.250} = 5.55.$$

Thus, observing data has updated our relative belief in the interactive model (against all other models) by a factor of 5.55. Notice that this is different from the Bayes factor in the column BF_{10}; this Bayes factor represents an update in our relative belief in the interactive model against the *null model only*.

Continuing these types of calculations will give us the full Bayesian ANOVA table, displayed here as Table 5.4. Notice that only the additive and interactive models have $\mathrm{BF}_{\mathcal{M}}$ greater than 1, indicating that these are the only two models that received an increase in model odds after observing data. Of these, the support for the interactive model is the largest.

With this example finished, we are now ready to perform a Bayesian ANOVA in JASP with our Tooth Growth dataset. If you've held off and not done it already, performing the test in JASP is as easy as you might expect. Click the "ANOVA" button and select the Bayesian ANOVA. The interface will look largely the same as in the classical ANOVA we did earlier. Move the len variable into the "Dependent Variable" box and move supp and dose into the "Fixed Factors" box. The model comparison table will

Table 5.4

Model	$p(\mathcal{M})$	$p(\mathcal{M} \mid \text{data})$	$\mathbf{BF}_{\mathcal{M}}$	$\mathbf{BF}_{10}$
Null	0.20	0.023	0.09	1
A	0.20	0.047	0.20	2
B	0.20	0.116	0.52	5
A+B	0.20	0.233	1.22	10
A+B+A*B	0.20	0.581	5.55	25

Figure 5.4 A JASP screenshot of a Bayesian ANOVA applied to the Tooth Growth dataset.

appear quickly in the results screen and is displayed in Figure 5.4. It is very important to note that the specific numbers you obtain will be slightly different. This is because the computation of the various Bayes factors involves approximation, which I'll say more about shortly.

Given the example we just walked through, this JASP output should now look familiar. There are two differences that I will point out. First, you'll notice that the order of the rows is different from that given in the previous example. That is because the default behavior of JASP is to order the models in terms of *decreasing* posterior probability. That is, it gives the best fitting model at the top. As a result, the column BF_{10} no longer expresses the Bayes factor for the given model over the *null* model, but rather as a Bayes factor for that model over the *best* model. You can change this display to one more closely aligned with our example by selecting the option "Compare to null model" under "Order" in the analysis pane (not that there is really any convincing reason to do so, in general).

The other difference is that there is an extra column ("error %"). The computation of each Bayes factor BF_{10} uses the method of

Rouder, Morey, Speckman, and Province (2012), which involves using Gaussian quadrature to approximate some integrals. Any kind of numerical approximation method involves some error, and this column quantifies this error. As we can see, the largest error we experience is about 5.3%, which is relatively small. In fact, the JASP team recommends trying to keep this error under 20% (see *van den Bergh et al., 2021*). If an error percentage above 20% is observed, one way to reduce this error is to increase the number of samples used in the numerical approximation. This setting can be adjusted in the "Additional Options" menu.

BRIDGING CLASSICAL AND BAYESIAN ANOVA

At its core, the Bayesian ANOVA in JASP is an exercise in model comparison. We can see from our model comparison table in Figure 5.4 that the preferred model is the interactive model, and given our descriptives plot, this is not too much of a surprise. However, reporting of results from an analysis of variance has classically proceeded as an assessment of whether each main effect and interaction is meaningful ("significant"), and the literature is full of examples of this type of reporting. So how do we construct a bridge between this type of "effects"-driven reporting and the model comparison approach we just described. The answer lies in something called the *inclusion Bayes factor*, which is part of a more general technique called *Bayesian model averaging*.

The idea is simple. If our goal is to assess whether the effects of each experimental manipulation are meaningful, one way to accomplish this in a Bayesian framework is to assess the evidence for *including* each main effect and interaction as a term in the model. The complicating issue is that there are several models being considered at once. A way around this complication is to do Bayesian model averaging (*Hinne, Gronau, van den Bergh, & Wagenmakers, 2020*). JASP implements Bayesian model averaging via the "Effects" option in the Bayesian ANOVA analysis pane. When you select "Effects", you'll get the second table (labeled "Analysis of Effects"). The result is shown in Figure 5.4.

Let's take a look at the information given to us in this table. Bayesian model averaging combines the evidence for *including* a particular experimental factor by averaging the evidence across all models which contain that factor. Here's how it works. The prior probability of including the factor supp in our model is 0.6; this is because three of the five models include supp, and each of these five models has a prior probability equal to 0.2. Similarly, the prior probability of including dose is also 0.6. Finally, the prior probability of including the interaction term supp*dose is 0.2 because only one of those five models includes the interaction.

After observing data, these prior probabilities are updated to posterior probabilities, displayed in the table in the column $p(\text{incl} \mid \text{data})$. Notice that the posterior probabilities of including each factor *increases*. As an example, consider the factor supp, whose posterior probability is now 0.995. This number comes from adding the posterior probabilities for the three models that contain the factor supp (i.e., $0.721 + 0.275 +$ something really small $\approx$ 0.995). Similarly, the posterior probabilities for including dose and the interaction can be computed. Converting these inclusion probabilities to inclusion odds (as we did earlier when we talked about the model Bayes factors BF_M), we can divide the posterior inclusion odds by the prior inclusion odds to get the *inclusion Bayes factor*. Including supp in the model produces an inclusion Bayes factor of $\text{BF}_{\text{inclusion}} = 144.048$. This means that the data have increased our prior odds for including supp as a predictor in our model by a factor of 144, which we interpret as very strong evidence for a main effect of supp. Note that dose has an extremely large inclusion Bayes factor, so the evidence for an effect of dose is extreme. Finally, we have positive evidence for the interaction between supp and dose in the form of an inclusion Bayes factor for the interaction of 10.320, indicating that our observed data is approximately ten times more likely under models including the interaction than under models not including the interaction.

HOW TO COMMUNICATE THE RESULTS

As with our previous chapters, it is important to have a discussion of how to communicate the results of our Bayesian ANOVA. Just like with the Bayesian correlation and t-test, your writeups should include both the model definitions as well as the results of the model comparisons. As estimation is a little more complicated here and typically not a primary output of factorial ANOVA, we omit it here.

Model definitions: As always, the first step of reporting any Bayesian analysis is to include a definition of the models being considered. Here is an example of what this could look like for the Bayesian ANOVA:

We analyzed the tooth growth data using a Bayesian analysis of variance (Rouder et al., 2012) in JASP. Mean odontoblast lengths (in microns) were submitted to a 2x2 between-subjects factorial design, with factors of supplement (VC or OJ) and dose (500, 1000, or 2000 micrograms). Assuming the principle of marginality, this resulted in five models to be compared: a null model, two single main effect models, a model including both factors as main effects (i.e., an additive model), and a full interaction model. All models were assumed to be equally likely, a priori, and thus prior model probabilities were each set to 0.20. Within each model, JASP defaults were used for the model parameters.

Model comparisons: Next, we should report the results of our model comparisons. There are two main things to describe here. First, we report the overall model comparison across the five models. Then, we report the inclusion Bayes factors that were obtained from Bayesian model averaging. Here is an example from our analysis:

Of the five models we compared, the model which best predicted the observed data was the full interaction model, which had a posterior model probability of 0.721. As we can see in Figure 5.4, the only two models for which the posterior odds were increased after observing data were the full interaction model and the additive model. The observed data updated the posterior odds for the interaction model by a factor of 10.32, whereas the posterior odds for the additive model (without the interaction) were updated by a factor of only 1.52. Compared to the additive model, the data were 2.62 times[1] more likely under the full interaction model.

From this model comparison, we additionally performed Bayesian model averaging to assess the evidence for including each factor as a predictor in the model. Figure 5.4 also shows the inclusion Bayes factors for each factor and the interaction, calculated taking the quotient of the posterior odds and prior odds for including each term. As we can see, there is a large amount of evidence for a main effect of supplement ($BF_{inclusion} = 144.05$*), indicating that the observed data are 144 times* *more likely under models including supplement as a factor compared to models not containing supplement as a factor. Similarly, there is extreme evidence for a main effect of dose ($BF_{inclusion} = 1.0 \times 10^{15}$). Finally, there is strong evidence for an interaction between supplement and dose ($BF_{inclusion} = 10.32$), indicating that the data are 10 times more likely under models containing the interaction compared to models not containing the interaction.*

POST HOC COMPARISONS

From this first pass through our Bayesian analysis of variance, we observed positive evidence for (1) a main effect of supplement, (2) a main effect of dose, and (3) an interaction between supplement and dose. Interpreting the first main effect is simple; since there are only two levels of supplement (OJ and VC), the only way to interpret this main effect is that the mean tooth length for guinea pigs who were administered orange juice (i.e., the OJ group) differed from those who were administered a vitamin C supplement (VC).

Further, we can examine the model-averaged posterior estimates to pinpoint this difference. In JASP, we can generate a table of model-averaged posterior estimates by selecting "Estimates" under "Tables" in the analysis screen. This will produce the posterior summary table displayed in Figure 5.4.

From the posterior summary table in Figure 5.4, we can construct of the mean for the OJ group by combining the grand mean (18.815) with the "effect" of OJ (1.681). This gives an estimated mean odontoblast length of $18.815 + 1.681 = 20.496$ microns. On the other hand, the mean odontoblast length for those in the VC group can be estimated $18.815 - 1.681 = 17.134$. Thus, we can confidently conclude that guinea pigs given orange juice had longer incisors (by $20.496 - 17.134 = 3.362$ microns) than did those given the vitamin C supplement.

However, our inference is not quite as clear cut for the main effect of dose because our dose variable had three levels: 500, 1000, and 2000. Thus, observing a main effect of dose indicates that there is a difference somewhere *among* these doses, but it is not clear from our modeling *where* the difference occurs. Perhaps the tooth length differs between those guinea pigs given doses of 500 and 1000, but those given a dose of 2000 do not differ from those given a dose of

1000. Perhaps the difference is between all three levels. The point is that there are several ways in which we could observe a main effect of dose – but simply knowing there is a main effect does not tell us which levels differ from each other. We need a way to further pin-point these differences among the condition means. This is where *post hoc comparisons* come in.

In its simplest version, a post hoc comparison can be performed by simply choosing two of the levels in the independent variable and comparing their means via a t-test. If we have three levels, then choosing two of these levels to compare can be done three ways: (1) group 1 versus group 2; (2) group 2 versus group 3; and (3) group 1 versus group 3. For the moment, let's just ignore Bayesian statistics and go back to our classical methods. One could proceed with a t-test for each of these three comparisons, but then the overall *Type 1 error rate* for each test (typically assumed to be $\alpha = 0.05$) would increase. Since we performed three tests, each with an error rate of 5%, the overall error rate would inflate to 15%. To compensate for this, one can adjust the error rate for each test by dividing by the number of tests performed. In this case, the error rate for each test would be $0.05 / 3 = 0.0167$ – performing this adjustment would make the overall error rate turn out to be 5%. This is called the *Bonferroni correction*, and it is a classical method for performing error control when making multiple post hoc comparisons. Note that it can also be applied directly to the observed p-value for each test. In this case, we would adjust the observed p-value by multiplying by the number of tests. The effect on inference would be the same – one would need to observe a p-value no larger than $p = 0.0167$ to have it count as "significant".

Against this background, how might one proceed with post hoc comparisons in a Bayesian context? Of course, there are multiple approaches, but the one employed in JASP is based on a combination of ideas dating back to *Jeffreys (1961)* and described further by *Westfall et al. (1997)*. Just like we see with the classical approach, a Bayesian post hoc comparison is done with a Bayesian t-test. But instead of adjusting the Type I error rate α or the observed p-value, the adjustment is performed on the prior odds of each pairwise difference. Then, we perform a Bayesian update by multiplying these (adjusted) prior odds by the observed Bayes factor for the pair-wise difference (computed via the Bayesian t-test we discussed in

Chapter 4). This gives the posterior odds in favor of the pairwise difference.

Now that we have a conceptual direction, let's go ahead and perform the post hoc comparison in JASP. After we see *what* we get from JASP, I'll spend some time describing *how* the computation is performed. To perform the post hoc comparison, you'll want to open the "Post Hoc Tests" menu in the analysis screen. We want to perform our post hoc comparisons on the *dose* variable, so highlight that variable and click the right arrow to move it into the right side pane (see Figure 5.5). Note that the "Correction" option "Null control" should be selected by default. This is the option that we want.

The output will immediately appear in the results screen under "Post Hoc Tests" (see Figure 5.5). Let's go through what this table is displaying. First, let's consider the third column, labeled "$BF_{10,U}$". This column displays the Bayes factor BF_{10} for the t-test comparing the two levels indicated in a given row. It is important to note that these Bayes factors are *uncorrected*; they are simply the Bayes factors one would obtain if a Bayesian independent samples t-test was performed on the two levels being compared.

The correction is applied by adjusting the prior odds for each of the comparisons and then using the obtained Bayes factor as an updating factor to compute posterior odds. Let's walk through one of those corrections. Consider the first row, which compares the observed odontoblast lengths len for doses of *500* versus *1000*. In this comparison, we pit the following models against each other: $\mathcal{H}_1 : \mu_{500} \neq \mu_{1000}$ versus $\mathcal{H}_0 : \mu_{500} = \mu_{1000}$, where μ_{500} and μ_{1000} are

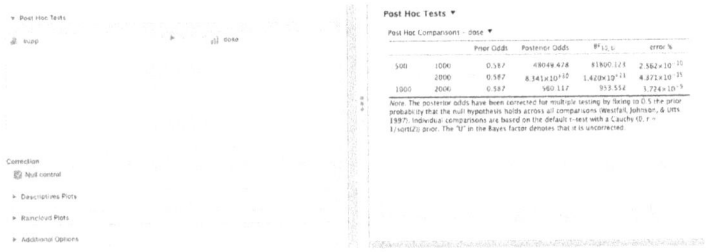

Figure 5.5 JASP screenshot of the Post Hoc Tests dialog in the Bayesian ANOVA analysis screen.

the means of the groups with doses of 500 and 1000, respectively. The default prior odds for the comparison in favor of $\mathcal{H}_1$ would be 1.0, meaning that the prior probabilities of $\mathcal{H}_1$ and $\mathcal{H}_0$ are equal (i.e., 1-to-1 odds). When we apply the correction, the prior odds in favor of $\mathcal{H}_1$ decrease to 0.587 (I'll explain how this number is computed momentarily). From here, we can use the Bayes factor to update these prior odds in favor of $\mathcal{H}_1$ into *posterior* odds for $\mathcal{H}_1$ by multiplying. For the first row, multiplying prior odds (0.587) by the observed (uncorrected) Bayes factor (81800.123) gives us posterior odds of 48049.478 – certainly large odds in favor of $\mathcal{H}_1$. In fact, we see quite large odds in favor of $\mathcal{H}_1$ for each of the pairwise comparisons, indicating that the mean odontoblast lengths are different for each of the three levels of dose.

Conceptually, what we are seeing is similar to a Bonferroni correction in the classical approach. In a Bonferroni correction, the diagnostic test statistic is the *p*-value, which is adjusted in such a way that its diagnostic value is attenuated. This is exactly what we are seeing here. By adjusting the prior odds downward from 1 to 0.587, our posterior odds in favor of $\mathcal{H}_1$ are decreased from what they would be if we just assumed the default of 1-to-1 prior odds. Thus, the diagnostic value of the Bayes factor in the follow-up *t*-test is attenuated to account for the multiple comparisons taking place. In this case, the attenuation does not qualitatively change the evidence categories we observe for each comparison. Indeed, each of the observed Bayes factors represents a large degree of evidence for $\mathcal{H}_1$, as do the attenuated posterior odds.

I'll conclude this section with an explanation of how the adjusted prior odds is computed. This little bit is math-heavy and can be skipped for now, but feel free to come back to it when you are ready. For simplicity, let's assume a scenario similar to our example where we have three condition means μ_1, μ_2, and μ_3. For a specific post hoc comparison (say μ_1 versus μ_2), our goal is to compute the prior odds of observing a difference (e.g., $\mu_1 \neq \mu_2$) compared to observing a null effect (e.g., $\mu_1 = \mu_2$). Thus, our goal is to compute the ratio

$$\frac{p\left(\mu_1 \neq \mu_2\right)}{p\left(\mu_1 = \mu_2\right)}.$$

The method described by *Westfall et al. (1997)* starts by assuming that the prior probability of $\mathcal{H}_0$ is 0.50. That is,

$$p(\mathcal{H}_0) = p(\mu_1 = \mu_2 = \mu_3) = 0.50.$$

Now, assume further that for each value of i ($i = 1,2,3$), we have $p(\mu_i = \mu) = \tau$, where μ is the grand mean. That is, a given condition mean is equal to the grand mean with probability τ, and drawn from some continuous distribution with probability $1 - \tau$. Because the distribution is continuous, two randomly drawn values are never exactly equal. Thus, if we know that μ_1 and μ_2 are equal, it must be the case that they are both equal to μ. This implies that we can write

$$p(\mu_1 = \mu_2) = p(\mu_1 = \mu) \times p(\mu_2 = \mu) = \tau^2.$$

By the same argument, we can also see that

$$p(\mathcal{H}_0) = p(\mu_1 = \mu_2 = \mu_3)$$
$$= p(\mu_1 = \mu) \times p(\mu_2 = \mu) \times p(\mu_3 = \mu) = \tau^3.$$

Solving for τ gives us

$$\tau = \sqrt[3]{p(\mathcal{H}_0)} = \sqrt[3]{0.50} = 0.7937.$$

Thus, the prior odds of observing a difference between condition means μ_1 and μ_2 can be calculated as:

$$\frac{p(\mu_1 \neq \mu_2)}{p(\mu_1 = \mu_2)} = \frac{1 - \tau^2}{\tau^2} = \frac{1 - 0.7937^2}{0.7937^2} = \frac{0.3700}{0.6300} = 0.587.$$

Notice that this is exactly the prior odds that JASP reports for each pairwise comparison in the Post Hoc Tests table.

NOTE

I will admit that it may be quite unsettling to the reader for me to seemingly gloss over the issue of defining priors for our *model parameters*. My rationale for this is twofold: (1) the priors used in the Bayesian ANOVA are well-considered (see *Rouder et al., 2012*) and are constructed to provide good objective performance; and (2) the priors are multivariate generalizations of the Cauchy prior we encountered in Chapter 4. Further, as my goal with this book is to provide the *basics* of Bayesian inference, I recognize that a detailed discussion of priors on parameters in the Bayesian ANOVA will necessarily go well beyond a "basic" level of exposition, so I leave those details to the interested reader to pursue in the primary literature (e.g., *Rouder et al., 2012, 2016*). Note that *Coon, Silva, Etz, and Sarnecka (2024)* provide a very accessible description of priors on parameters in Bayesian ANOVA models. Also, readers interested in applying Bayesian methods to *repeated-measures* designs should consult *van den Bergh, Wagenmakers, and Aust (2023)* for an excellent tutorial on performing such a test in JASP.

CHAPTER SUMMARY

Let's recap what we've learned in this chapter:

1. We introduced the Bayesian analysis of variance, which works by comparing multiple models of some continuous outcome variable with various combinations of categorical predictors and their interactions.
2. We walked through the calculation of a model comparison table, which JASP uses to communicate the results of the Bayesian analysis of variance.
3. We described how to use Bayesian model averaging to bridge the Bayesian concept of model comparison with the classical "effects"-driven reporting in classical ANOVA. The main tool of interest is the *inclusion Bayes factor*, which represents the factor by which the observed data are more likely when including a predictor in the model.
4. We described a procedure for conducting Bayesian post hoc comparisons in analysis of variance. The method works by adjusting the prior odds for a specific pairwise difference downward, which then attenuates the value of the observed Bayes factor.

EXERCISES

1. A study investigated the effectiveness of small-group tutor-
    ing and the effectiveness of a classroom instruction technique
    known as *hot math* (*Fuchs et al., 2008*). The hot math program
    teaches students to recognize types or categories of problems
    so that they can generalize skills from one problem to another.
    The data in Table 5.5 is a math test score for each student after
    16 weeks in the study. You can either input this into JASP
    yourself, or you can load the CSV file which you can down-
    load from https://t.ly/aJPcY.

Table 5.5

	No tutoring	With tutoring
Traditional instruction	3,6,2,2,4,7	10,4,5,8,4,6
Hot math instruction	7,7,2,6,8,6	8,12,9,13,9,9

a. Sketch a plot of the condition means implied by this
    experimental design. Plot *instruction type* on the horizontal
    axis and separate lines for each level of *tutoring group*.
b. Perform a classical analysis of variance on these data.
    Report *p*-values for the main effects of *instruction type* and
    tutoring group and the interaction.
c. Perform a Bayesian analysis of variance on these data.
    Which model has the largest posterior probability? Com-
    pute and report inclusion Bayes factors for each main
    effect and the interaction.
d. Re-do the model averaging, but this time select "Across
    matched models" rather than the default "Across all mod-
    els". What changes? What do you think is going on here?
e. Given your analyses, do you think the effectiveness of *hot
    math* instruction depends on whether tutoring is available?
    Explain.

NOTE

1   This factor can be calculated by dividing the BF_{10} values for each model.
    From Figure 5.4, this calculation is 1.00/0.381 = 2.62.

BAYESIAN LINEAR REGRESSION

Earlier in this book, we walked through examples of performing Bayesian inference for simple situations involving measuring the degree of association via correlation tests (Chapter 3) and comparing group means via a t-test (Chapter 4). This work gave us a bedrock of knowledge that is fundamental to Bayesian inference, including concepts such as posterior distributions and Bayes factors, both of which provide us with easy-to-interpret information. Then, we extended these simple tests to the analysis of variance, allowing us to test the effects of experimental manipulations on some continuous outcome variable. When we have manipulations encoded by multiple independent variables, this results in a set of several models that need to be compared. Additionally, we encounter multiple levels of uncertainty, both between models as well as within models. In Chapter 5, we introduced Bayesian model averaging as a tool for managing these multiple sources of uncertainty.

In this chapter, we end our set of "case studies" by considering a generalization of the ANOVA framework we started in Chapter 5. Here, we will consider the situation where our goal is to predict some continuous outcome variable as a function of several predictor variables (which may be continuous themselves). Fortunately, most of the work we did in Chapter 5 will serve us well in this chapter.

DOI: 10.4324/9781003470199-6

EXAMPLE: TEACHING STATISTICS DURING A PANDEMIC[1]

When the COVID-19 pandemic began to shut down universities in the spring semester of 2020, faculty and administrators needed to figure out how to quickly adjust our traditional methods of instruction to allow for maximum flexibility. After all, most university students (and faculty) were sent home to "shelter in place" during early 2020, and when classes returned after that summer, there were many restrictions on how many people could be gathered in one place at one time. This meant that faculty had to think critically about how they can best deliver instruction in new formats. Indeed, it was the wild west, and in hindsight, it was a critical moment in the history of education. Ever the optimist, I decided to make the best of the situation and repackage my experience with teaching statistics during a pandemic into an example of how to do Bayesian linear regression.

Before jumping into the example, I need to provide some background. At my university, we opted to follow the "HyFlex" model of instruction, where instructors continued to teach their courses in a face-to-face format, but the lectures were simultaneously streamed online and recorded. This gave students three options for attendance – they could flexibly choose whether to attend (1) face-to-face; (2) remote synchronous; or (3) remote asynchronous. With the last two options, students "attended" the course from a remote location, but they still had to choose whether to log in and participate during the scheduled time of lecture (i.e., synchronous attendance) or watch the pre-recorded lecture at a different time (asynchronous attendance).

As one might imagine, this new freedom of choice that was afforded to thousands of our university students meant that our lecture halls quickly became quite empty. There were many days that I taught my class to *one* student! While a few intrepid souls regularly attended their face-to-face classes (proudly wearing their masks and/or plastic face shields), many opted to attend remotely. Perhaps unsurprisingly, even more of our students opted for asynchronous remote attendance. After all, if you can watch the lecture whenever you want, why watch at 8:00 in the morning? Quickly, our faculty and administration picked up on this pattern and began to notice that students weren't performing as well as they should, especially among our students who opted for asynchronous attendance. This observation led many faculty members and department leaders to think seriously about removing the asynchronous option as a possibility for future semesters.

Well, not so fast. Maybe there's more to the story than first meets the eye. With this in mind, I decided to collect some data first. My specific aim was to answer the following question: Do my students' course grades depend on whether they attend lectures synchronously or asynchronously?

Okay, so let's talk about the data. While it is not available in the JASP Data Library, you can easily access it from the following web address: https://osf.io/yf2sb/. The file statGrades.csv contains some anonymous course performance data from 33 students in my first-year statistics course. I included the final course grade (on a scale of 100 points) for each student. I also categorized each student as a synchronous student or an asynchronous student. I did this by counting the number of lectures attended by each student throughout the semester. Any student who attended at least 75% of lectures (either in person or remotely) was categorized as synchronous (sync = 1), and everyone else was categorized as asynchronous (sync = 0).

To load these data into JASP, you'll first want to download the file statGrades.csv somewhere on your computer. Then in JASP, click "Open", navigate to the "Computer" option, and click the "Browse" button. There, you'll open a file system dialog where you'll locate the file you just downloaded.

As a first step, let's take a quick look at some descriptive statistics. Based on what I've already described, there is probably quite a bit of difference in the course grades between synchronous and asynchronous attenders (that's certainly the impression that our faculty and administration seemed to have at the time). Indeed, that seems to be the case with these data. If you click the "Descriptives" button, move grade to the "Variables" list, and split by sync, we get the table shown in Figure 6.1.

As we can see, there is just over a 15-point advantage for the synchronous attenders (sync = 1) compared to the asynchronous

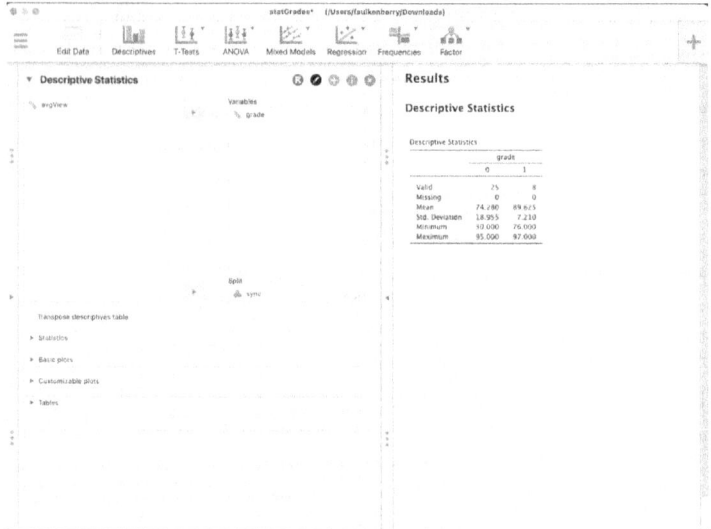

Figure 6.1 A JASP screenshot of the descriptive statistics for `grade`, split by `sync`.

attenders (`sync` = 0). But look at those standard deviations; there is quite a bit more variation among the asynchronous attenders, so clearly something else is going on.

Fortunately, I had some additional data that might explain some of this variability. In conversations with my students during that Fall semester class, it became clear that some of my asynchronous students were not actually watching the recorded lecture videos. Since video viewing times for each student were available from our learning management system, I was able to figure out how many minutes of each recorded lecture video that each student watched. From these data, I computed the average length of time that each student watched the lectures during the semester. This mean (standardized to a maximum of 75 minutes – the length of one class session) is recorded in the variable `avgView`.

Thus, we now have two important variables that might contribute to some of the variability in course grades. One way to better understand this relationship is to perform a Bayesian linear regression, which we can easily do in JASP.

PERFORMING A BAYESIAN LINEAR REGRESSION

Linear regression is a general statistical technique that can be used to assess the specific impacts of a set of predictor variables on some outcome variable. Since I ultimately would like to know the extent to which sync and avgView predict course grade, linear regression will be a good tool for this. Further, *Bayesian* linear regression lets us answer this question by integrating hypothesis testing and estimation into a single analysis.

Fortunately, we've already covered a lot of the background needed to understand Bayesian linear regression in Chapter 5, where we described the Bayesian analysis of variance. So much of this will seem familiar to you. Our two predictors sync and avgView give us a total of four[2] models that we can test against our observed data. Once we've chosen the best model (i.e., the one that best predicts the observed data), we can then use the models to estimate the impact of each predictor.

Let's now describe the four models. We'll start with the most complex (i.e., the one with both predictors contributing to the model) and work our way down the list in order of decreasing complexity.

- Model 1: grade ~ sync + avgView

    This model hypothesizes that a student's course grade is impacted both by their attendance (synchronous versus asynchronous) AND the average amount of time that the student spent watching the lectures.
- Model 2: grade ~ sync

    Compared to Model 1, this model drops average viewing time as a predictor and, thus, hypothesizes that course grade is impacted by attendance mode but NOT by the average lecture viewing time.
- Model 3: grade ~ avgView

    Compared to Model 1, this model drops attendance mode as a predictor and, thus, hypothesizes that course grade is impacted by average lecture viewing time but NOT by attendance mode.
- Model 4: Null model

    This model hypothesizes that neither attendance mode nor average lecture viewing time predicts course grade.

As in all of our previous examples of Bayesian analyses, our first task is to determine which of these models is best supported by the observed data. In JASP, we click on the "Regression" button and select "Bayesian Linear Regression". Next, move `grade` into the "Dependent Variable" box, and move the two predictor variables `sync` and `avgView` into the "Covariates" box. If you cannot move `sync` into the box, it is likely because JASP has imported the column sync as a categorical variable. You'll want to switch back to the data view and change the variable type to "Scale" – you do this by clicking on the little symbol beside the variable name in the data sheet. If it has been imported as a categorical variable, the symbol will look like a Venn diagram. Just click on that symbol and select the Scale option, which is depicted as a ruler.

One additional thing you'll want to do before proceeding is to go to the "Advanced Options" menu and change the "Model Prior" from "Beta binomial" to "Uniform". I'll explain what this is a bit later.

The first output that JASP produces is our model comparison table,[3] which will look somewhat similar to the table produced when we did Bayesian ANOVA in Chapter 5. The table is displayed in Figure 6.2. The model comparison table tells us which of the four models displays the best predictive adequacy – that is, which

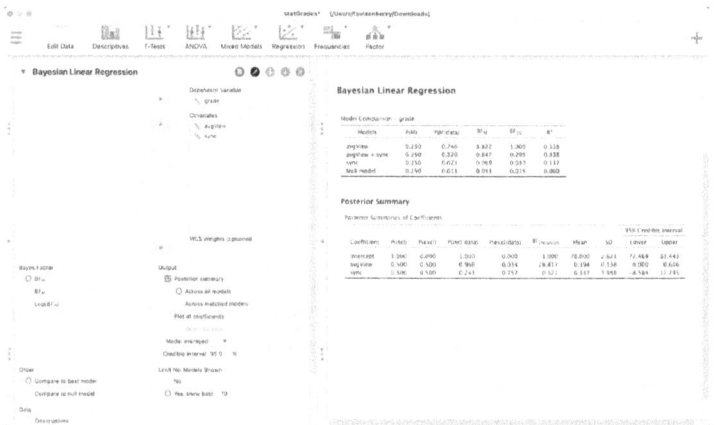

Figure 6.2 Screenshot of a Bayesian linear regression performed in JASP.

model does the best job of predicting the observed data. By default, the models are listed in order from most predictive to least predictive. From the model comparison table in Figure 6.2, we can see immediately that our data are most likely under the model containing only average viewing time as a predictor. Let's take a closer look at why this is the case.

The first two columns are $p(\mathcal{M})$ and $p(\mathcal{M} \mid \text{data})$. $p(\mathcal{M})$ denotes the prior probability of each model. Since we chose "Uniform" under "Model Prior" in the advanced options, each of these models is assumed to be equally likely before observing data. Given that there are four models under consideration, a uniform prior on models would make the prior probability of each model equal to $1/4 = 0.25$. The column labeled $p(\mathcal{M} \mid \text{data})$ contains the posterior probability of each model – that is, after observing data. The best fitting model, containing only avgView as a predictor, has a posterior probability of 0.746, whereas the next best fitting model (sync + avgView) has a posterior probability of 0.220. The remaining two models account for a combined posterior probability of $0.023 + 0.011 = 0.034$. In other words, these two models are not very likely at all.

The next two columns are $\text{BF}_{\mathcal{M}}$ and BF_{10}. As we discussed in Chapter 5, these are both Bayes factors, but they are slightly different types of Bayes factors. $\text{BF}_{\mathcal{M}}$ is a Bayes factor on the *model odds* – that is, it is the factor by which the odds in favor of a specific model change after observing data. Observing $\text{BF}_{\mathcal{M}} > 1$ means that the model's odds have *increased* after seeing the data, whereas observing $\text{BF}_{\mathcal{M}} < 1$ means that the model's odds have *decreased*. Let's work through an example to make this a bit clearer. Consider the best fitting model, containing only one predictor, avgView. Before observing data, the odds in favor of this model are 1-to-3, or 0.333. We can see this by dividing the prior probability of the model (0.250) by the probability of all other models (0.250 + 0.250 + 0.250 = 0.750) – that is, $0.250/0.750 = 0.333$ (after all, "odds" is just a ratio of two probabilities). How do these odds shift after observing data? In a similar manner, let's compute the posterior odds for the avgView model: $0.746/(0.220 + 0.023 + 0.011) = 2.937$. If we divide these posterior odds (2.937) by the prior odds (0.333), we get the updating factor of $\text{BF}_{\mathcal{M}} = 8.822$. We interpret this number in the following way: *after observing data, our odds in favor of the model containing only average viewing time as a predictor have increased by a factor of 8.822.*

On the other hand, BF_{10} gives the relative predictive adequacy of the given model compared to the best fitting model. Notice that the second-best fitting model (`sync + avgView`) has $BF_{10} = 0.295$. This means that the observed data are 0.295 times as likely to occur under this two-predictor model than they are under the single-predictor `avgView` model. Taking reciprocals ($1/0.295 = 3.389$), we can interpret this a bit more easily: *the observed data are 3.389 times more likely under the model containing only average viewing time as a predictor compared to the model that also specifies whether the student is a synchronous or asynchronous attender.*

So what does this all mean? One conclusion we may immediately draw is that average lecture viewing time is clearly a predictor of course grade, because the posterior probability of including it in the model is $0.746 + 0.220 = 0.966$. Now our question shifts to the following: Does it matter whether a student attends synchronously or asynchronously? To answer this question, we need compare the model containing both predictors to the model containing only average viewing time – we can use both of our obtained Bayes factors to make this comparison. First, only `avgView` received increased support after observing data (the model odds were increased by a factor of 8.822) – all other models received decreased support. Additionally, compared to `sync + avgView` (where attendance mode matters), the data are 3.389 times more likely under the single predictor model `avgView`. So does attendance mode matter? Given these data, I would argue that the answer is no. Instead, it is the average lecture viewing time that best predicts course grades.

Now that we've established that average lecture viewing time matters, the next step is to estimate its impact. *How much gain in course grade can I expect for each additional minute of average viewing time?* To answer this, we need to look at the next outputs in JASP.

First, let's consider the posterior summary table, which we can select as an option in our analysis pane. The posterior summary table provides information about each possible predictor in the linear regression model and is shown in Figure 6.2.

Roughly, the posterior summary table consists of two parts. The first part (including all columns to the left of and including the column $BF_{inclusion}$) helps us determine *whether* to include each possible predictor in the model. The second part (including the remaining columns to the right) tells us about the coefficients of each predictor. But there is so much more going on here – and it all deals with uncertainty. Let's look deeper.

Recall from our earlier discussion of the model comparison table that we have uncertainty about which model best predicts our observed data. Certainly, we believe that the model with the single predictor of avgView is best, but there is also a small probability that the two-predictor model is the right one. Since my goal is to inform my own future policy about permitting asynchronous attendance in my courses, I would like to know which predictors I should include in the model. JASP helps answer this using Bayesian model averaging, which combines the evidence for including a particular predictor by averaging across the models that contain that predictor. Here's how it works. The prior probability of including the variable sync in our model is 0.5; this is because two of the four models include sync. Similarly, the prior probability of including avgView is also 0.5. After observing data, these prior probabilities are updated to posterior probabilities. The posterior probability of including sync now falls to 0.243; this number comes from adding the posterior probabilities for the two models that contain the variable sync (i.e., 0.220 + 0.023 = 0.243). Equivalently, we can also say that the posterior probability of *excluding* sync from our model is 1–0.243 = 0.757.

On the other hand, the posterior probability of including avgView increases to 0.966. Converting these inclusion probabilities to inclusion odds (as we did earlier when we talked about the model Bayes factors BF_M), we can divide the posterior inclusion odds by the prior inclusion odds to get the inclusion Bayes factor. Including avgView in the model produces an inclusion Bayes factor of $BF_{inclusion}$ = 28.817. This means that the data have increased our prior odds for including avgView as a predictor by a factor of 28.817, which we interpret as strong evidence for including avgView in the model. In other words, average viewing time matters! On the other hand, the data have *decreased* our prior odds for including sync. We can take the reciprocal of the inclusion Bayes factor to get an *exclusion* Bayes factor:

$$\mathrm{BF}_{\text{exclusion}} = \frac{1}{\mathrm{BF}_{\text{inclusion}}} = \frac{1}{0.321} = 3.12\,.$$

Thus, my observed data is 3.12 times more likely under models *not* including sync as a predictor. Based on this positive evidence for excluding sync, I will choose to only include average viewing time as a predictor of course grade (and leave out attendance mode).

So what does this have to do with *estimating* the impact of average viewing time? As we know from our work in the previous chapters, a Bayesian analysis provides not only a point estimate for each predictor's coefficient (the column labeled "Mean"), but it also captures the uncertainty around this point estimate via a 95% credible interval. However, it is important to note that any estimate we make is *conditional on the underlying model*. For example, the estimate of the effect of avgView will be different under the single-predictor model than under the two-predictor model that also includes sync.

Thus, we have uncertainty in two places. There is uncertainty in the estimate itself, and there is also uncertainty in the model choice. Bayesian model averaging provides an elegant way to account for *both* sources of uncertainty. The 95% credible intervals that we see for each coefficient in Figure 6.2 reflect a weighted average where each estimate is weighted by the posterior probability of including that specific predictor in the model. Thus, the resulting credible intervals account not only for uncertainty within the model, but also uncertainty across the models. This averaging becomes apparent when we look at the marginal posterior distribution plots in Figure 6.3.

From the posterior summary table in Figure 6.3 we can see that the coefficient of avgView has a posterior mean of 0.394. This means each additional minute of watching the recorded lecture videos improves course grade by an average of 0.394 points. Said differently, every additional 25 minutes of average viewing time improves course grade by 10 points (a "letter grade" in the US grading system). The model-averaged credible interval tells us that this coefficient is 95% probable to be between 0.075 and 0.683. Notice the small "spike" at 0 on the left tail of the top plot in Figure 6.3. This spike reflects the (albeit small) probability of 0.0335 of excluding avgView as a predictor.

On the other hand, consider the marginal posterior distribution for the coefficient of sync. Even though the posterior summary

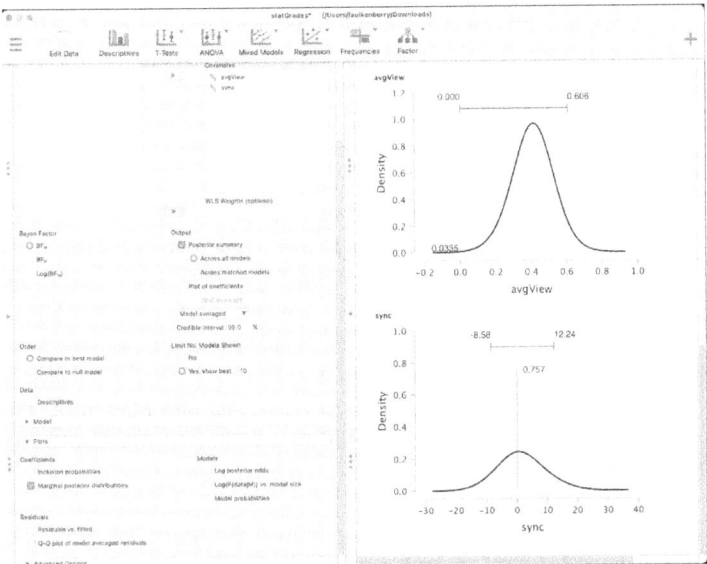

Figure 6.3 A plot of the marginal posterior distributions for the coefficients of
avgView and sync.

table in Figure 6.2 gives us an estimate for this coefficient, the plot
of marginal posterior distribution for the coefficient of sync (bot-
tom plot in Figure 6.3) shows a large spike at 0. This spike reflects
the large probability (0.757) of excluding sync as a predictor in the
model. If sync is included in the model (the probability of including it is 0.243), it is 95% probable that the effect of synchronous
attendance is between −7.64 points and +12.62 points. Clearly, we
do not see a consistent effect of synchronous attendance. In fact, in
our preceding discussion we argue that there is positive evidence for
the *absence* of any such effect.

HOW TO COMMUNICATE THE RESULTS

In many ways, the mechanics of Bayesian linear regression mirror
those of the Bayesian analysis of variance from Chapter 5. Thus, the
way we will write up the results for this Bayesian linear regression
will be similar to that of the Bayesian ANOVA.

Model definitions: Time and time again throughout the book, we have emphasized that the first step of reporting any Bayesian analysis is to include a definition of the models being considered. That continues to be the case here. Here is an example of what this could look like for our Bayesian linear regression:

We analyzed the course grade data using a Bayesian linear regression (van den Bergh et al., 2021) in JASP. There were four models under consideration: a null model, two single-predictor models where final course grade was predicted solely by average viewing time or attendance mode, and an additive model which contained both average viewing time and attendance mode as predictors. The four models under consideration were assumed to be equally likely, a priori, and thus prior model probabilities were each set to 0.25. Within each model, JASP defaults were used for the model parameters.

Model comparisons: Next, we report the results of our model comparisons. There are two main things to describe here. First, we report the overall model comparison across the four models. Then, we report the inclusion Bayes factors that were obtained from Bayesian model averaging. Here is an example from our analysis:

Of the four models we compared, the model which best predicted the observed data was the model which contained a single predictor of average viewing time, which had a posterior model probability of 0.746. As we can see in the model comparison table shown in Figure 6.2, the only model for which the posterior odds were increased after observing data were this single predictor model. The observed data updated the posterior odds for this single predictor model by a factor of 8.8. All other models saw their posterior odds decreased after observing data. Compared to the additive model also including attendance mode as a predictor, the data were 3.39 times[4] more likely under the single predictor model containing only average viewing time as a predictor.

From this model comparison, we additionally performed Bayesian model averaging to assess the evidence for including each predictor in the model. The posterior summary table in Figure 6.2 shows the inclusion Bayes factors for each possible predictor, calculated by taking the quotient of the posterior odds and prior odds for including each term. As we can see, there is strong evidence for including average viewing time as *a predictor (* $BF_{inclusion} = 28.8$ *), indicating that the observed data are almost 30 times more likely under models including average viewing time as a predictor compared to models not including average viewing time as a predictor. However, the inclusion Bayes factor for attendance mode is less than 1, indicating that the data are more likely if we exclude attendance mode as a predictor. Specifically, we have positive evidence (* $BF_{exclusion} = 3.12$ *) for excluding attendance mode as a predictor.*

AN ALTERNATIVE WAY TO ASSIGN MODEL PRIORS

While many aspects of the Bayesian linear regression procedure in JASP mirror those of the Bayesian analysis of variance we covered in Chapter 5, there is one glaring difference – the default specification on the prior probability of the competing models is different. In fact, you may have found it odd that we had to go into the "Advanced Options" menu to select the uniform model prior. This is because JASP uses by default something called a *beta binomial* prior for model probabilities. In this section, we will describe this beta binomial prior and discuss why it is used as the default model prior specification for Bayesian linear regression in JASP. Additionally, we will re-do our analysis using this beta binomial prior and see to what extent our conclusions are robust against changing the prior model probabilities.

Before introducing the beta binomial distribution as a solution to some unknown problem, we need to identify the *problem* first. Consider the working example that we've used in this chapter. Here, we have built a regression model to predict grade with two possible predictors: avgView and sync. As we've seen, the Bayesian approach is to build four models: the null model, the two single-predictor models, and the additive two-predictor model. The way I've described this decomposition of the "model space" indicates that there are three *classes* of models that are built: (1) models with zero predictors, (2) models with one predictor, and (3) models with two predictors.

Using the uniform prior on models, then, results in a bit of a conundrum. While each of the four models we built is equally likely, a priori, this prior specification actually results in a non-uniform prior on model *classes*. To see why this is the case, consider the class of models with exactly one predictor. There are two of these models: grade ~ avgView and grade ~ sync. Since each has a prior probability of $p(\mathcal{M}) = 0.25$, the prior model probability on the class of single-predictor models is $0.25 + 0.25 = 0.50$. This means that our prior belief is *biased* toward single-predictor models – in this case, single-predictor models are twice as likely, a priori as all other models.

A solution to this problem is to implement a two-stage, hierarchical approach to assigning prior model probability (*Scott & Berger, 2006, 2010*). The idea is to first assign a uniform prior to model *classes* (i.e., models with zero predictors, models with one predictor, etc.) rather than the models themselves. Then, we assign an equal probability to all models within each class. The diagram in Figure 6.4 illustrates this process for a regression model with two predictor variables. First, we have three model classes: (1) models with zero predictors, (2) models with one predictor, and (3) models with two predictors. Since there are three such classes, a uniform prior probability for each class would be equal to $1/3 = 0.333$. Now we consider a next step, shown schematically in Figure 6.4.

For each of these model classes, we consider how many models are contained within each class. For the class of models with zero predictors, there is exactly one model – the null model. Dividing the prior probability of the class (0.333) by the number of models (1) gives a prior probability for the null model of $0.333/1 = 0.333$. The same idea works for the class of models with two predictors, as there is only one model in this class (the full, two-predictor model grade

Model classes:	0 predictors	1 predictors	2 predictors
Prior probability of class:	0.333	0.333	0.333
Number of models:	1	2	1
Prior probability each model *within class*:	0.333	0.167	0.333

Figure 6.4 A schematic diagram showing how the beta binomial prior assigns prior probabilities to each model in a Bayesian linear regression.

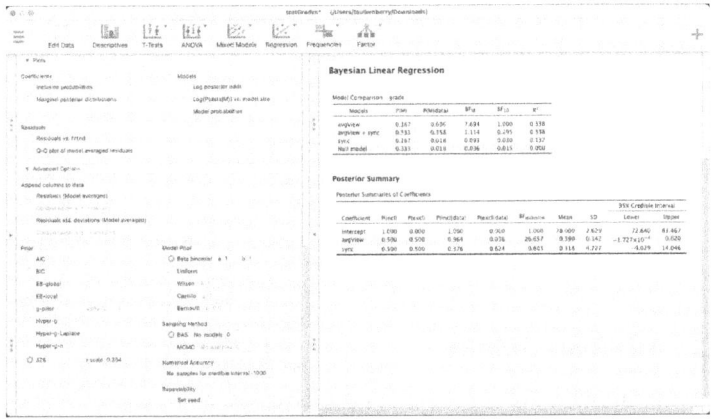

Figure 6.5 A JASP screenshot of the output obtained when using the beta binomial prior on models.

~ avgView + sync). For the class of models with one predictor, there are two such models. Dividing the prior probability for the class (0.333) by the number of models (2) gives a prior probability for each of the single-predictor models equal to $0.333/2 = 0.167$.

To see this in action in JASP, you'll want to go back to the Advanced Options menu and select the "Beta binomial" model prior. All the previously selected outputs in JASP will dynamically update to reflect this new model prior specification.

As we can see in Figure 6.5, the model comparison table has changed slightly from that given under the uniform model prior specification. First, note that the columns BF_{10} and R^2 did not change. This is because these columns do not depend on the prior model probability. The remaining columns have changed a bit. Qualitatively, however, there are no major differences between this table and our previous table. We see that the single predictor model avgView is the only one that substantially increases its model odds. Whereas the two-predictor model avgView + sync saw its model odds decrease ($\text{BF}_{\mathcal{M}} = 0.847$) when we specified a uniform model prior, its model odds slightly increase ($\text{BF}_{\mathcal{M}} = 1.114$) under the beta binomial model prior. This is entirely because this model's prior weight in the overall model space has increased, and the resulting updating through data has increased its posterior odds just slightly.

However, this support is weak, so the qualitative conclusions do not change in a substantive manner.

In fact, looking at the posterior summary table in Figure 6.5 justifies this conclusion. Indeed, even under a different model prior specification, the odds of including avgView as a predictor in our model have increased after seeing data ($BF_{inclusion} = 26.657$), and our odds of including sync have decreased ($BF_{inclusion} = 0.603$). This is exactly the same pattern of results we observed under the original uniform model prior.

CHAPTER SUMMARY

Let's recap what we've learned in this chapter:

1. We introduced the Bayesian linear regression, which works in a similar manner to the Bayesian analysis of variance. The approach works by comparing multiple models of some continuous outcome variable with various linear combinations of predictors (which may be continuous).
2. We walked through performing the Bayesian linear regression in JASP and how to interpret the model comparison table as well as interpreting the strength of predictors via inclusion Bayes factors.
3. We discussed how Bayesian model averaging can be used to estimate the posterior distribution of a predictor's slope (i.e., effect) by simultaneously incorporating within-model and across-model uncertainty.
4. We described a different approach to assigning prior model probabilities called the *beta binomial* prior, which assigns prior model probabilities in two stages. First, different model classes containing each number of possible predictors are assigned uniform model probabilities, then each different model within each class is assigned a uniform probability within that class.

EXERCISES

1. *Holloway and Ansari (2009)* examined the relationship between basic numerical processing and higher mathematical skills in children. In their study, they examined whether children's

mathematical achievement scores were related to their ability to compare the sizes of symbolic numbers (i.e., comparing numbers presented as digits) and nonsymbolic numbers (i.e., comparing arrays of dots). For each of the children in their study, they measured mathematical fluency scores (encoded in the dataset as fluency) as well as standardized measures of performance in both symbolic and nonsymbolic number comparison tasks (encoded, respectively, as symbolicDE and nonsymbolicDE). You can download a copy of their observed data as a CSV file from https://t.ly/SkvvW.

a. Perform a Bayesian linear regression on these data, using fluency as an outcome variable, and symbolicDE and nonsymbolicDE as predictor variables. Change the model prior to uniform for this analysis.

b. Which model has the largest posterior probability? Which models increase their posterior odds after observing data?

c. Use Bayesian model averaging to assess the evidence for including (or excluding) each predictor.

d. Re-do the analysis using the beta binomial prior. How does your output change?

e. Given your analyses, which type of number magnitude knowledge (symbolic or nonsymbolic) seems to impact mathematical fluency?

NOTES

1   I posted an earlier version of this chapter in the JASP blog in November 2020. The original post can be found at https://jasp-stats.org/2020/11/26/how-to-do-bayesian-linear-regression-in-jasp-a-case-study-on-teaching-statistics/.

2   By default, JASP does not include interaction terms in linear regression models, though they can be added if desired (van den Bergh et al., 2021).

3   Just like in Chapter 5, your numbers may vary slightly from what is depicted in Figure 6.2. This is because the Bayes factors are computed using numerical approximations.

4   This factor can be calculated by dividing the BF_{10} values for each model. From Figure 6.2, this calculation is $1.00/0.295 = 3.39$.

7

NEXT STEPS AND FURTHER READING

When you first started reading this book, you likely had some motivation to learn the basics of Bayesian statistics. After reading the last six chapters, you have begun to get a good idea of *what* Bayesian statistics is, *why* it is a good approach for statistical inference, and importantly, *how* it works. Further, you have gained some experience with JASP, a free and easy-to-use software package for doing statistical inference. Importantly, I hope you have finished this book with a desire to learn more about Bayesian statistics. In this chapter, I will briefly recap where we've been on our journey through the basics of Bayesian statistics, and I will give you some suggestions for the next steps you might take in your future Bayesian journey.

A BRIEF RECAP

We began our journey in Chapter 1, which served as an introduction to the core principles of statistical inference. There, we focused our efforts on understanding the classical null hypothesis testing framework, which relies on *p*-values to index support for a null hypothesis about some population parameter. Recall that the *p*-value represents the likelihood of observing some data (or more extreme) under the null hypothesis. Thus, a small *p*-value means that the data we observed should be *rare* if the null hypothesis were true. Observing a small *p*-value leads us to *reject* the null hypothesis as a viable model for the underlying population, giving us indirect evidence for an alternative hypothesis. We conclude this chapter by critically examining some limitations of *p*-values, including their dependence on data that were never observed, their dependence

DOI: 10.4324/9781003470199-7

on researchers' subjective intentions, and, critically, their inability to quantify statistical evidence. This discussion then set the stage for our introduction of Bayesian methods as a better tool for our statistical inference.

Chapter 2 was a long chapter, but a very important one for us. There, we concentrated on learning the essential terms and foundational concepts of Bayesian statistics, with a strong emphasis on the ideas of *prior* and *posterior* distributions. The chapter began with a discussion of Bayes' theorem, a reasonably simple mathematical equation that can be used to describe exactly how one updates prior beliefs after observing new data. One of the key mechanical components of Bayes' theorem is the prior distribution, which we use to give an initial representation of our uncertainty about parameters before observing any data. We then demonstrate how this prior distribution is transformed into a posterior distribution, both intuitively and visually through the JASP Learn Bayes module, as well as mathematically.

We also introduced the Bayes factor as a model comparison tool in Chapter 2. The Bayes factor has two conceptual definitions that are both useful in a variety of contexts. One way to define a Bayes factor comes from the Bayesian concept of a prior predictive distribution – here, the Bayes factor is considered as the relative likelihood of the observed data under both competing models. For example, a Bayes factor of $BF_{10} = 10$ would mean that the data are ten times more likely under the alternative hypothesis than under the null hypothesis. Another way to define the Bayes factor comes from Bayes' theorem itself – in this case, the Bayes factor is considered as the factor by which the prior odds between two competing models is updated to posterior odds. This is a useful idea that allows us to compute the posterior probability of a hypothesis after observing data.

By the end of Chapter 2, you should have gained an appreciation for how Bayesian methods allow for a more intuitive understanding of statistical uncertainty. One key advantage over classical inference is the ability of Bayesian statistics to provide us with a measure of strength of evidence for either hypothesis, not just the null. By the end of this chapter, you will have gained a solid foundation in the language and principles of Bayesian statistics, preparing you for the practical applications we explored in the following chapters.

Chapters 3 and 4 began our "case study" chapters, each of which allowed us to explore the use of Bayesian statistics with real data. Chapter 3 considered correlation, where we measured the strength of association between various components of the Big Five personality dimensions. Chapter 4 considered the t-test, which we used to assess group differences in performance on a stereogram task. Both chapters used real data from scientific studies, all of which is available in the JASP Data Library. Another important feature of these chapters was the detailed guidance on how to communicate the results of our Bayesian analyses.

In addition to applying the concepts we learned in Chapter 2, each of these chapters provided something extra to extend our knowledge of Bayesian statistics. In Chapter 3, we introduced the notion of a sensitivity plot, which allows us to explore the impact of changing prior specifications on our test parameter. In Chapter 4, we added a discussion of how to perform Bayesian tests of *directional* hypotheses, focusing on both intuitive as well as detailed descriptions of the underlying mathematics.

After gaining confidence with the basic Bayesian hypothesis tests of Chapters 3 and 4, we moved on to more complex tests in Chapters 5 and 6. In Chapter 5, we introduced the Bayesian analysis of variance (ANOVA), emphasizing its importance as a foundational tool for analyzing experimental data. The chapter began by outlining the classical ANOVA approach, which decomposes total variance into components attributable to different sources, such as treatments and error. I used the classic "Tooth Growth" dataset as a working example to illustrate a two-factor factorial design, where the effects of vitamin C supplementation (in two forms) and dosage levels (three levels) on tooth length were examined. This set the stage for understanding the importance of main effects and interactions in experimental designs.

We then transitioned to Bayesian ANOVA, highlighting how this method evaluates the predictive performance of multiple competing models simultaneously. We also learned about Bayesian model averaging, and in particular, the inclusion Bayes factor, which quantifies the evidence for the presence (or absence) of specific main effects and interactions in the model. Using Bayesian model averaging to assess the evidence for main effects and interactions provides us with a bridge between classical and Bayesian approaches to ANOVA.

In Chapter 6, we shifted our focus to Bayesian linear regression, extending the discussions from previous chapters to provide a comprehensive understanding of how to predict continuous outcomes based on multiple predictor variables. I illustrated the application of this framework in the context of teaching statistics during the COVID-19 pandemic, using real data to highlight the practical implications of Bayesian methods. A key focus of the chapter is on addressing model uncertainty in our parameter estimates through Bayesian model averaging. I explained how this technique helps to account for uncertainty in model selection by combining predictions from multiple models, providing a more robust estimate of the coefficients of our regression model.

In the end, we are now equipped with the knowledge and skills to effectively conduct Bayesian analyses. By emphasizing the practical application of these techniques, we showcased the flexibility and interpretative power of Bayesian statistics in understanding complex relationships in real-world data. I am hopeful that you now feel empowered to apply Bayesian methods confidently in your own research endeavors!

FURTHER BAYESIAN EXPLORATION

After working through this book, you might be interested in knowing about other Bayesian methods that can be applied to other types of research questions. There are so many to consider, but I will focus on a few emerging techniques that are included in JASP. The descriptions that follow will just scratch the surface, but I will provide references to guide you in case you are interested in applying these methods in your own work.

Bayesian re-analyses of summary statistics

Consider the following situation: you are reading a paper on some phenomenon that is of interest to you, and the authors claim support for a *null* effect. While you see that they have given the relevant test statistic – for example, $t(48) = 1.32$, $p = 0.19$ – you do not have access to the raw data. Since you know from Chapter 1 that observing a p-value greater than 0.05 does *not* index the actual evidence for the null hypothesis, you wonder if it is possible to get a Bayes factor directly from this *summary*. As it turns out, the answer is yes!

The answer lies in the JASP Summary Statistics module, which is one of the add-on modules that is included with every JASP download. The basic idea of the Summary Statistics module is that you can provide the minimal summary statistics that are provided in a paper (usually, the test statistic and some information about sample size), and JASP will produce a Bayes factor directly from this minimal input. For example, with the data summary presented in the previous paragraph, the JASP Summary Statistics module will produce a Bayes factor of $BF_{01} = 1.739$, which we interpret as only anecdotal evidence for the null hypothesis. This can be a useful tool when reading papers on a topic because it allows you to gauge the actual evidence provided by someone's claimed results. In this case, we might not be too confident in this reported result and choose to study it further.

A good tutorial paper for exploring Bayesian reanalysis of summary statistics is the following:

- *Ly, A., Raj, A., Marsman, M., Etz, A., & Wagenmakers, E.-J. (2018)*. Bayesian reanalyses from summary statistics: A guide for academic consumers. *Advances in Methods and Practices in Psychological Science, 3,* 367–374. doi: 10.1177/2515245918779348

Note that as of the writing of this book, the JASP Summary Statistics module works only with t-tests, correlation, linear regression, and tests of proportions. However, I have several papers on performing Bayesian re-analyses of ANOVA summaries (e.g., *Faulkenberry, 2021; Faulkenberry & Brennan, 2023*), and those methods are used in a simple online calculator called PsyStat, which can be accessed from any device at https://tomfaulkenberry.shinyapps.io/psystat.

Bayesian tests of proportions

When we first learned about Bayesian methods in Chapter 2, we walked through an example of a binomial test, which is a simple test of one proportion against a hypothesized value. In that example, we were interested in knowing whether my university's proportion of "green chile" people was different from 0.5. However, what if you want to test *two* sample proportions against each other? Certainly, this is something that is very common in clinical research. For example, in pharmacological studies, one group of study participants might be given a certain drug, whereas the other group is given a placebo. Then, suppose each group is measured on some binary outcome variable, such as whether they experienced a side effect or not. In this case, the critical test is whether the proportion of participants who experienced a side effect differs between the drug group and the placebo group.

For situations like these, an ideal test is the Bayesian A/B test, which is implemented in JASP. The test works by converting success proportions to odds ratios. Then, the "effect" is measured by building a linear model on the logarithm of this odds ratio. Thus, the prior and posterior are on one model parameter (the log odds ratio), making it easy to interpret the output in a manner similar to our previous one-parameter examples of Bayesian correlation and the Bayesian t-test.

While the details of this test are beyond the scope of this book, you can walk through some examples in the following tutorial paper:

- *Hoffmann, T., Hofman, A., & Wagenmakers, E.-J. (2022)*. A tutorial on Bayesian inference for the A/B test with R and JASP. *Methodology*, *18*, 239–277. doi: 10.5964/meth.9263

Bayesian meta-analysis

The goal of a meta-analysis is to statistically combine results that are observed across several studies. This usually amounts to using the sample statistics reported in papers to estimate the mean and standard deviation of the underlying *population* effect (*Field & Gillett, 2010*). The key questions in meta-analysis, then, are ones of presence and variability. That is, is there evidence for an overall effect, and is there variability among these effects?

To answer these questions, one then builds four models to be considered. These four models are formed by considering the possible answers to the aforementioned two questions. To make this clear, we need to introduce some mathematical notation. Let δ_i represent the true effect size for the i-th study and let μ represent the overall population effect size. Then, we can define our four models as follows:

- Model 1: The overall population effect size μ is 0, and each study effect size δ_i is also 0. This is called a *fixed-effects null hypothesis*.
- Model 2: The overall population effect size μ is 0, but the study effect sizes δ_i do have some variability. This is called a *random-effects null hypothesis*.
- Model 3: The overall population effect size μ is nonzero, and there is no variability among the study effects δ_i (that is, $\delta_i = \mu$ for all *i*). This is called a *fixed-effects alternative hypothesis*.
- Model 4: The overall population effect size μ is nonzero, and there is some variability among the study effects δ_i. This is called a *random-effects alternative hypothesis*.

Classically, one first assesses whether the effects are variable, and then after committing to either this fixed- or random-effects model, tests the null hypothesis that the overall effect is 0. This approach has several downsides. Certainly, there are the limitations of classical null hypothesis testing that we described in Chapter 1, but this concern is compounded by the uncertainty related to our choice of whether the effect is fixed or random.

A Bayesian solution to this problem is to incorporate model uncertainty by using Bayesian *model averaging* (*Gronau, Heck, Berkhout, Haaf, & Wagenmakers, 2021*), which we described in Chapters 5 and 6. Considering the uncertainty about the four models simultaneously, we can index the overall evidence for the null versus the alternative hypothesis as well as the overall evidence for the existence versus absence of variability among the effects. And, like our previously described Bayesian methods, this approach gives the researcher parameter estimates for the population mean and standard deviation, and rather than being conditional on a given model, these estimates are "unconditional", formed by averaging the posteriors across the four models and weighting each model by its plausibility.

This Bayesian meta-analysis method is implemented in JASP as one of the extra modules that may be activated on demand. Again, as the details are beyond the scope of this book, I will invite you to consider the following excellent tutorial:

- *Berkhout, S. W., Haaf, J. M., Gronau, Q. F., Heck, D. W., & Wagenmakers, E.-J. (2023)*. A tutorial on Bayesian model-averaged meta-analysis in JASP. *Behavior Research Methods, 56*(3), 1260–1282. doi: 10.3758/s13428-023-02093-6

FURTHER READING FOR GENERAL KNOWLEDGE

Now that you know the basics of Bayesian statistics, I will leave you with a list of five books and five papers that I personally feel are some of the best resources to extend your Bayesian journey after learning the basics. Please note that for the most part, these readings will be quite advanced, but you are prepared to begin exploring them. And don't worry if you don't understand everything when you read it the first time. Indeed, it will take many years to develop a deep knowledge of a subject. The advantage you have now, after reading this book, is that you know *the basics,* and that is exactly the ticket you need to begin learning the deep, fascinating world of Bayesian statistics.

First, I will share some books that I think are good ones to consider buying. Indeed, these five books have a prominent place on my own bookshelf.

- *Bayesian Data Analysis, 3rd Edition (2013)* by Andrew Gelman, John B. Carlin, Hal S. Stern, David B. Dunson, Aki Vehtari, and Donald B. Rubin, published by CRC Press.
  - *Bayesian Data Analysis* has become a classic textbook for Bayesian inference. If you can only afford one book, make it this one! Its depth and range of topics is unparalleled, and scholars in Bayesian statistics (including myself!) frequently come back to this book for reference. Its topic coverage is staggering, consisting of 23 chapters and 3 appendices over 675 pages. In their book, Gelman et al. cover Bayesian methods with an applied focus, covering advanced topics such as hierarchical models, model checking, and computational techniques like Markov Chain Monte Carlo (MCMC). It emphasizes real-world applications, offering insights on regression models and updates on modern methods such as Hamiltonian Monte Carlo and nonparametric modeling.

- *Statistical Rethinking: A Bayesian Course with Examples in R and STAN* – 2nd Edition (*2020*) by Richard McElreath, published by CRC Press.
  - If you can only afford one book, get the Gelman et al. book. However, if you can afford two, *Statistical Rethinking* should be your next purchase! Compared to *Bayesian Data Analysis*, this book is more conceptual and less formal in its mathematical descriptions. However, it is still aimed at readers with more experience (i.e., graduate students and professionals) and covers Bayesian statistics in considerable depth. If you enjoyed my book, you will definitely get a lot out of this one. *Statistical Rethinking* integrates R and the Stan programming language, providing readers with a deep toolkit of computational tools for applying Bayesian methods.
- *A Student's Guide to Bayesian Statistics* (*2018*) by Ben Lambert, published by Sage Publications.
  - Lambert's *A Student's Guide to Bayesian Statistics* is marketed as the "first student-focused introduction to Bayesian statistics". It certainly achieves this goal in terms of the friendly visual formatting of the book and the intuitive narrative throughout, but similar to the two competitors previously mentioned, it also covers a very large range of topics. I think it is a good bridge between my book and the Gelman et al. and McElreath books. If you find Gelman et al. and McElreath to be too daunting, consider reading *A Student's Guide* next!
- *Practical Bayesian Inference: A Primer for Physical Scientists* (*2017*) by Coryn Bailer-Jones, published by Cambridge University Press.
  - Don't let the title fool you – *Practical Bayesian Inference* is not just for physical scientists. This is another good textbook that covers the essential ideas of Bayesian statistics, including lots of hands-on examples using the statistical programming language R.
- *Bayesian Cognitive Modeling: A Practical Course* (*2014*) by Michael Lee and E.J. Wagenmakers, published by Cambridge University Press.
  - I saved my favorite for last. While it is geared toward cognitive scientists, *Bayesian Cognitive Modeling* gives the reader

a quick exposure to the key components of Bayesian statistics through a problem-based learning approach. The book is more like a workbook than a traditional textbook. Instead of just reading, you are invited to work along with the examples and play around with the R code. You'll learn a lot doing this, as I did.

- *Bayesian Inference from the Ground Up: The Theory of Common Sense* (*2024*) by E.J. Wagenmakers and Dora Matzke, freely downloadable from https://bayesianspectacles.org/free-course-book.
  - I know this is an *extra* book, but since it is free to download, just consider this as a bonus. And what a bonus it is! In this enjoyable and comprehensive (and free!) text, you will literally learn Bayesian inference "from the ground up". Integrating hands-on examples in JASP with lots of history and philosophy, this book will provide you with hours of entertaining reading. If any part of my book has left you still wondering "why", you'll likely find it in *Bayesian Inference from the Ground Up*.

Finally, I will share some of the papers that I think are essential reading for an aspiring Bayesian. Of course, these will be mostly focused on my own research field of the behavioral sciences, but I think the knowledge contained within will nicely translate to your own field of interest. This was a difficult list to construct, but what remains are my five favorite papers to recommend to my own students. I hope you enjoy them too!

- *Wagenmakers, E.-J.* (*2007*). A practical solution to the pervasive problems of *p* values. *Psychonomic Bulletin & Review*, *14*(5), 779–804. doi: 10.3758/BF03194105
  - This was one of the first papers that turned me on to the idea of using Bayesian statistics in my own research, and I have come back to it many times over the past 17 years. In this paper, Wagenmakers argues for the superiority of Bayesian inference over classical methods and shows how Bayesian methods can solve many of the problems associated with traditional null hypothesis testing, such as overconfidence in *p*-values and lack of replicability. This paper has had a major influence on the ongoing debate regarding

methodological reform in psychology, advocating for the widespread use of Bayesian methods in place of or alongside classical approaches.

- Rouder, J. N., Speckman, P. L., Sun, D., Morey, R. D., & Iverson, G. (*2009*). Bayesian t-tests for accepting and rejecting the null hypothesis. *Psychonomic Bulletin & Review*, *16*(2), 225–237. doi: 10.3758/PBR.16.2.225

  - Alongside the Wagenmakers paper, this paper by Jeff Rouder and colleagues was one of my foundational readings when I was initially learning about Bayesian methods. Although it is focused on the development of the Bayesian *t*-test that we covered in Chapter 4, its discussion of the *what* and *why* of Bayesian inference is very good. As I've told Jeff many times, I get something new from this paper every time I read it. This is because his papers have so much depth to them. Read it once, then read it again . . . I guarantee you'll find something new every time.

- Masson, M. E. J. (*2011*). A tutorial on a practical Bayesian alternative to null hypothesis significance testing. *Behavior Research Methods*, *43*(3), 679–690. doi: 10.3758/s13428-010-0049-5

  - Though the Wagenmakers and Rouder papers were published first, it is Mike Masson's paper referenced here that finally made it "click" for me. Masson's tutorial shows how to effectively use traditional statistical inference as a bridge to Bayesian inference through something called the *BIC* approximation. In fact, after I worked through this paper, I had an idea while out on a run one morning that allowed me to bypass some of Masson's calculations and compute Bayes factors directly from the summary statistics presented in papers. This paper (Faulkenberry, 2018) was my first significant contribution to the field of Bayesian statistics. In some respects, I owe my career to Mike Masson and this paper . . . I doubt he knows this.

- van Doorn, J., van den Bergh, D., Böhm, U., Dablander, F., Derks, K., Draws, T., . . . Wagenmakers, E.-J. (*2021*). The JASP guidelines for conducting and reporting a Bayesian analysis. *Psychonomic Bulletin & Review*, *28*(3), 813–826. doi: 10.3758/s13423-020-01798-5

- These last two papers are somewhat recent and represent really well-written tutorial papers on doing Bayesian inference in JASP. This first one, led by Johnny van Doorn, uses the stereogram example we covered in Chapter 4. But, while we focused on the basics of the t-test in our chapter, van Doorn and colleagues use this example to comprehensively cover all aspects of conducting and reporting a Bayesian analysis.
- *Coon, J., Silva, P. N., Etz, A., & Sarnecka, B. W. (2024).* Bayesian tools of the trade for developmental psychologists: A quickstart guide using JASP. *Journal of Cognition and Development.* doi: 10.1080/15248372.2024.2386032
  - This is a paper that I am really excited about! It just came out as I was finishing this book, and I have already started recommending it to people as one of the best single tutorial papers on Bayesian inference that I have ever read. While it is focused on developmental psychology, its descriptions of the underlying concepts and the detailed examples of writeups are unparalleled, both in terms of depth and clarity. You should immediately add this one to your reading list.

GLOSSARY

alternative hypothesis – see **alternative model**.

alternative model – a statistical model (typically denoted $\mathcal{H}_1$) that states that the population mean is different from some fixed number. Usually used to represent a hypothetical situation where a treatment population mean is different from the mean of the general population. Also called an **alternative hypothesis**.

analysis of variance – a technique for performing statistical inference on a collection of three or more sample means. Instead of testing models of the population mean, the analysis of variance works by partitioning variance of observed data into two sources – the variance between treatments and the variance within treatments – and computes the ratio of these two variances. Inference is performed by comparing this ratio against a standard F-**distribution**.

ANOVA – see **analysis of variance**.

Bayes factor – a number used for model comparison in Bayesian inference, it represents the relative likelihood of some observed data under two different models. It also represents the factor by which the prior odds in favor of some model is increased after observing data (i.e., it is the ratio of posterior odds to prior odds).

Bayesian model averaging – a method that accounts for uncertainty in model selection by averaging over possible models, weighting each by its posterior probability given the data, to make more robust inferences about possible parameter values.

Bayesian model comparison – a method of model comparison that relies on the Bayes factor instead of the p-value to index how well a set of competing models predicts some observed data.

Bayes' Rule – see **Bayes' Theorem**.

Bayes' Theorem – also known as **Bayes' Rule**, it is a mathematical result that formalizes the computation of posterior probability as the prior probability multiplied by an updating factor.

beta binomial distribution – in the context of a model prior for Bayesian linear regression, it is a hierarchical distribution where the number of predictors is modeled as a binomial distribution, and the models within each model class are distributed according to a beta distribution (by default, a uniform distribution).

beta distribution – in the context of Bayesian inference, it is a continuous probability distribution used to represent prior beliefs about the probability of success in a binomial model, with its shape determined by two parameters that reflect the nature of prior information.

binomial distribution – a probability distribution that models the number of successes in a fixed number of independent trials, where each trial has two possible outcomes (success or failure) and the same probability of success (see also **binomial model**).

binomial model – in the context of Bayesian statistics, it describes the likelihood of observing a certain number of successes in a fixed number of independent trials as a function of the probability of success on each trial, incorporating prior beliefs about this probability of success and updating these beliefs with observed data to form a posterior distribution.

Bonferroni correction – a statistical adjustment used to reduce the likelihood of Type I errors when performing multiple comparisons, typically applied by dividing the significance level by the number of tests conducted.

Cauchy distribution – in the context of Bayesian inference, it is a heavy-tailed probability distribution often used as a prior for scale parameters due to its ability to allow for greater uncertainty and accommodate extreme values. It is particularly useful as a prior for effect size in the Bayesian t-test.

conditional probability – the probability of an event occurring, given that another event has already occurred.

correlation coefficient – a descriptive statistic that describes the degree of association between two numeric variables. Ranges from -1 to 1.

credible interval – in the context of Bayesian inference, it is a range of values within which an unknown parameter is likely to lie, with a specified probability (typically 95%), based on the posterior distribution.

descriptive statistics – numerical summaries of a set of measurements. Typically examples include mean, median, mode, variance, and standard deviation.

directional – in the context of research questions or hypotheses, it represents a question of whether a treatment mean changes in a specific direction (i.e., either increases or decreases).

effect size – a measure of the impact of an experimental treatment, defined in terms of the observed difference between measurements divided by the standard deviation.

estimation – in the context of Bayesian inference, it is the process of summarizing the posterior distribution to gain information about some unknown population parameter.

exclusion Bayes factor – quantifies the strength of evidence for excluding a particular predictor or effect in a model by comparing models with and without that predictor based on how well they explain the observed data.

frequentist – a theoretical perspective on probability which specifies that probability measures represent the relative long-run frequency of events. For example, a frequentist interpretation of a p-value of 0.05 would consider that the given observed data (or more extreme) would only occur 5 times out of 100 if the null model were correct.

Gaussian distribution – see **normal distribution**.

hierarchical model – is a structured model that incorporates multiple levels of parameters, where parameters at one level are governed by distributions that depend on parameters at higher levels.

hypothesis testing (also known as **model comparison**) – the process of assessing which model is most likely to predict some observed data. Hypothesis testing traditionally involves comparing two models against a set of observed data.

inclusion Bayes factor – quantifies the strength of evidence for including a particular predictor or effect in a model by comparing models with and without that predictor based on how well they explain the observed data.

independent samples _t_-test – a specific type of _t_-test used for contexts where the goal is to compare the means of two independent, or nonoverlapping, samples.

interaction effect – in the context of experimental design, it occurs when the effect of one factor on the outcome depends on the level of another factor, indicating that the factors do not act independently on the outcome variable.

inverse chi-square distribution – in the context of Bayesian inference, it is a probability distribution commonly used as a prior for variance parameters in hierarchical models, usually for representing uncertainty about the variance of a normal distribution.

linear regression – a method of statistical inference where outcome variables can be mathematically predicted from predictor variables.

main effect – in the context of experimental design, it refers to the independent impact of a single factor or variable on the outcome of interest, averaged across all levels of other factors in the experiment.

marginal distribution – typically used in the context of Bayesian inference, it is the probability distribution of one (or more) variable(s) that is obtained by integrating or summing over the possible values of the other variables in a joint distribution.

marginality principle – it states that models should include the main effects of factors whenever interaction effects involving those factors are present, ensuring that the hierarchical structure of effects is properly respected. Also known as the Principle of Marginality.

marginal probability (see also **marginal distribution**).

mean – also known as the average, it is a measure of center that is defined as the sum of the measurements divided by the number of measurements.

median – for a finite dataset, it is defined as the middle number. For a continuous distribution, it is defined as the value for which 50% of the distribution lies below that value.

model – a quantitative instantiation of some observable phenomenon. In this book, models are usually _statistical_ models, so they are specified as values of parameters according to some probability distribution.

model comparison (also known as **hypothesis testing**) – the process of assessing which model is most likely to predict some observed data. Whereas hypothesis testing is traditionally done in the context of two models, model comparison can involve considering any number of candidate models against a set of observed data.

nondirectional – in the context of research questions, it represents a question of whether a treatment mean changes, without any specific prediction of direction for the change.

normal distribution – also known as a **Gaussian distribution**, it is one of the most common probability distributions used in statistical inference. Though it has a specific mathematical definition, it is usually described as mound shaped and symmetric about its center. It is completely described by two parameters: the mean μ and the standard deviation σ.

null hypothesis – see **null model**.

null hypothesis testing – a classical form of model comparison which is based on computing the probability of observing some specific data (or more extreme) if the null hypothesis is true.

null model – a statistical model (typically denoted $\mathcal{H}_0$) that states that the population mean is equal to some fixed number. Usually used to represent a hypothetical situation where a treatment population mean is no different from the mean of the general population. Also called a **null hypothesis**.

odds – an alternative way to express uncertainty, it is defined as the ratio of two probabilities. Often used in the context of prior and posterior model odds (i.e., $\mathcal{H}_0$ versus $\mathcal{H}_1$).

one-tailed test – a model comparison or hypothesis test in which the alternative model is directional. In the context of classical hypothesis testing, is called "one-tailed" because the p-value comes from only one tail of the comparison distribution.

parameter – a number which specifies some essential component of a probability distribution. They are typically denoted by a Greek letter. For example, in the normal distribution, the center of the distribution is specified by the parameter μ, and the variability of the distribution is specified by the parameter σ.

paired-samples t-test – a specific type of t-test used in repeated-measures designs. Instead of computing the mean score among a single set of measurements, the observed scores are calculated as the difference between the two repeated measurements for each experimental unit.

population – in the context of a research question, this is the set of all observable units to which we want to generalize an observable phenomenon.

posterior distribution – in the context of Bayesian inference, it is the probability distribution of a model's parameter values *after* observing data.

posterior model probability – in the context of Bayesian inference, it is the probability distribution that mathematically encodes our belief about the possible values of a model's parameters *after* observing data.

post hoc comparison – a statistical analysis conducted after an ANOVA to explore specific group differences when a meaningful effect has been found, helping to identify which particular groups differ from each other.

principle of marginality – see **marginality principle**.

prior distribution – in the context of Bayesian inference, it is the probability distribution that mathematically encodes our belief about the possible values of a model's parameters *before* observing data.

prior model probability – in the context of Bayesian inference, it is the probability of a specific model *before* observing data.

prior predictive distribution – in the context of Bayesian inference, it is the probability distribution of observed data generated by integrating over all possible values of the parameters according to the prior distribution, before considering any actual data. It can be used to assess the performance of a model against actually observed data.

probability density function – see **probability distribution**.

probability distribution – a mathematical function (also called a **probability density function**) which formally expresses the likelihood of all possible measurements/outcomes.

***p*-value** – a number that expresses the probability of obtaining an observed sample (or more extreme) if the null model is correct. Used in classical hypothesis testing, it indexes the plausibility of observed data under the null.

sample – a subset of a population that is actually observed/measured.

standard deviation – a measure of variability, defined as the square root of **variance**. Compared to variance, standard deviation has the advantage of being on the same scale as the original measurements.

standard normal distribution – a normal distribution with $\mu = 0$ and $\sigma = 1$.

statistical inference – the process of using parameter estimation and model comparison to justify population-level claims about observable phenomena using observed sample data.

statistically significant – used in the context of classical statistical inference, it is a descriptive phrase that implies that an observed relationship would be very rare if there was indeed no relationship at the population level.

stretched beta distribution – used as a default prior for correlation coefficients, it is a symmetric beta distribution of the form Beta(a,a) with domain transformed from [0,1] to [−1,1]. Usually described in terms of *width,* defined as $1/a$.

student's *t*-distribution – see *t*-distribution.

***t*-distribution** – also known as **student's *t*-distribution**, it is a model for the distribution of sample means that is obtained whenever the population standard deviation σ is estimated from observed data. In simple cases, the t-distribution has one parameter – the **degrees of freedom** – and this parameter controls the shape of the distribution.

***t*-test** – a model comparison procedure used when the population standard deviation σ is unknown. In this case, σ is estimated from data, and the resulting distribution of sample means follows a ***t*-distribution** rather than a normal distribution.

two-tailed test – a model comparison or hypothesis test in which the alternative model is nondirectional. In the context of classical hypothesis testing, it is called "two-tailed" because the p-value comes from both tails of the comparison distribution.

uniform distribution – a probability distribution where all possible outcomes are equally likely.

variability – in the context of descriptive statistics, a single number which represents the extent to which the measurements in a set of data differ from the most typical measurement. Examples include standard deviation and variance.

variance – a measure of variability, defined as the average squared deviation from the mean. Compared to standard deviation, the variance is not as commonly used as a descriptive measure of variability because the scale is squared compared to the original measurement units.

ANSWERS TO SELECTED END-OF-CHAPTER EXERCISES

Chapter 1

1a. $\mathcal{H}_0 : \rho = 0$ and $\mathcal{H}_1 : \rho \neq 0$.

1b. $r = 0.055, p = 0.223$

1c. Since $p > 0.05$, we conclude that the observed data is plausible under the null hypothesis. Thus, we do not reject $\mathcal{H}_0$.

1d. 95% confidence interval = $[-0.033, 0.142]$

1e. The population correlation between extraversion and agreeableness is not significantly different from 0.

1f. The p-value is cut in half. This is because the p-value is calculated from only one tail instead of two.

2. Math SAT is significantly correlated with college GPA, $r = 0.252$, $p < 0.001$. However, verbal SAT is not significantly correlated with college GPA, $r = 0.114$, $p = 0.087$. Thus, math SAT is the better predictor of college success.

Chapter 2

1c. They all peak to the right of 0.5 but have varying levels of spread.

1d. For models 1, 2, and 3, the Bayes factors are 2.14, 0.68, and 3.75, respectively.

1e. The Bayes factors are quite different and depend heavily on the choice of prior.

2a. $\pi\left(\theta = 0.5\right) = 1$, and $\pi(\theta = 0.5 \mid \text{data}) = 7$.

2b. The observed data is seven times more likely under $\mathcal{H}_0$ than it is under $\mathcal{H}_1$.

Chapter 3

1a. There is a positive relationship between Openness and Agreeableness

1b. The default prior is a uniform distribution on ρ.

1c. The observed Bayes factor is $BF_{10} = 32.6$, indicating that the data are over 32 times more likely under $\mathcal{H}_1$ than under $\mathcal{H}_0$.

1d. Posterior median = 0.158, 95% credible interval = [0.072, 0.243].

1e. Perform a sensitivity analysis and notice that the observed Bayes factor ranges between 10 and 100 across a wide range of reasonable values for the stretched beta prior width.

2a. $p = 0.048$, so we reject the null and conclude that there is a statistically significant correlation.

2b. $BF_{01} = 1.51$, indicating that the data are 1.5 times more likely under the null.

Chapter 4

1c. $BF_{-0} = 4.33$, indicating that the data are 4.3 times more likely under the alternative.

1d. For most reasonable values of the Cauchy prior scale, the data provides moderate evidence for claimed effect.

2a. $BF_{10} = 1.32$, indicating that the data are 1.32 times more likely under the alternative. This is only anecdotal evidence for Borota et al.'s claimed effect.

2b. For Cauchy prior scales above 1, the data are more likely under the null.

Chapter 5

1b. Main effect of instruction is statistically significant, $F(1,20) = 10.996$, $p = 0.003$. Main effect of tutoring is also statistically significant, $F(1,20) = 12.289$, $p = 0.002$. Interaction is not significant, $F(1,20) = 1.086$, $p = 0.310$.

1c. The additive model has the largest posterior probability, $P(M \mid data) = 0.556$. For the main effect of instruction, $BF_{inclusion} = 10.65$. For the main effect of tutoring, $BF_{inclusion} = 14.50$. For the interaction, $BF_{inclusion} = 2.15$.

1d. The inclusion Bayes factors for the main effects are roughly the same, but the inclusion Bayes factor for the interaction drops below 1. Note: this better matches the results of the classical ANOVA in 1b.

Chapter 6

1b. The model containing only `symbolicDE` as a predictor has the largest posterior probability, $P(M \mid data) = 0.704$. This model and the additive model increase their posterior odds after observing data.

1c. There is strong evidence for including `symbolicDE`, $BF_{inclusion} = 138.27$, and there is anecdotal evidence for excluding `nonsymbolicDE`, $BF_{exclusion} = 2.44$.

1d. When using the beta binomial prior on models, the posterior probabilities and inclusion/exclusion Bayes factors change slightly, but the qualitative conclusions remain the same.

REFERENCES

Bailer-Jones, C. (2017). *Practical Bayesian inference: A primer for physical scientists.* Cambridge University Press.

Berkhout, S. W., Haaf, J. M., Gronau, Q. F., Heck, D. W., & Wagenmakers, E.-J. (2023). A tutorial on Bayesian model-averaged meta-analysis in JASP. *Behavior Research Methods*, *56*(3), 1260–1282. doi: 10.3758/s13428-023-02093-6

Birnbaum, A. (1962). On the foundations of statistical inference. *Journal of the American Statistical Association*, *53*, 259–326. doi: 10.1080/01621459.1962.10480660

Bliss, C. I. (1952). *The statistics of bioassay*. New York: Academic Press Inc.

Borota, D., Murray, E., Keceli, G., Chang, A., Watabe, J. M., Ly, M., . . . Yassa, M. A. (2014). Post-study caffeine administration enhances memory consolidation in humans. *Nature Neuroscience*, *17*(2), 201–203. doi: 10.1038/nn.3623

Cohen, J. (1988). *Statistical power analysis for the behavioral sciences* (2nd ed.). Hillsdale, NJ: Lawrence Erlbaum Associates.

Coon, J., Silva, P. N., Etz, A., & Sarnecka, B. W. (2024). Bayesian tools of the trade for developmental psychologists: A quick-start guide using JASP. *Journal of Cognition and Development*. doi: 10.1080/15248372.2024.2386032

Costa, P. T., & McCrae, R. R. (1992). Normal personality assessment in clinical practice: The NEO personality inventory. *Psychological Assessment*, *4*(1), 5–13. doi: 10.1037/1040-3590.4.1.5

Crampton, E. W. (1947). The growth of the odontoblasts of the incisor tooth as a criterion of the vitamin C intake of the guinea pig. *The Journal of Nutrition*, *33*(5), 491–504. doi: 10.1093/jn/33.5.491

Dolan, C. V., Oort, F. J., Stoel, R. D., & Wicherts, J. M. (2009). Testing measurement invariance in the target rotated multigroup exploratory factor model. *Structural Equation Modeling: A Multidisciplinary Journal*, *16*, 295–314. doi: 10.1080/10705510902751416

Faulkenberry, T. J. (2018). Computing Bayes factors to measure evidence from experiments: An extension of the BIC approximation. *Biometrical Letters*, *55*, 31–43. doi: 10.2478/bile-2018-0003

Faulkenberry, T. J. (2021). The Pearson Bayes factor: An analytic formula for computing evidential value from minimal summary statistics. *Biometrical Letters*, *58*(1), 1–26. doi: 10.2478/bile-2021-0001

Faulkenberry, T. J. (2022). *Bayesian statistics: The basics*. Routledge. doi: 10.4324/9781003181828

Faulkenberry, T. J., & Brennan, K. B. (2023). Computing analytic Bayes factors from summary statistics in repeated-measures designs. *Biometrical Letters*, *60*(1), 1–21. doi: 10.2478/bile-2023-0001

Field, A. P., & Gillett, R. (2010). How to do a meta-analysis. *British Journal of Mathematical and Statistical Psychology*, *63*(3), 665–694. doi: 10.1348/000711010x502733

Frisby, J. P., & Clatworthy, J. L. (1975). Learning to see complex random-dot stereograms. *Perception*, *4*(2), 173–178. doi: 10.1068/p040173

Fuchs, L. S., Fuchs, D., Craddock, C., Hollenbeck, K. N., Hamlett, C. L., & Schatschneider, C. (2008). Effects of small-group tutoring with and without validated classroom instruction on at-risk students' math problem solving: Are two tiers of prevention better than one? *Journal of Educational Psychology*, *100*(3), 491–509. doi: 10.1037/0022-0663.100.3.491

Gelman, A., Carlin, J. B., Stern, H. S., Dunson, D. B., Vehtari, A., & Rubin, D. B. (2013). *Bayesian data analysis*. Chapman and Hall/CRC. doi: 10.1201/b16018

Gronau, Q. F., Heck, D. W., Berkhout, S. W., Haaf, J. M., & Wagenmakers, E.-J. (2021). A primer on Bayesian model-averaged meta-analysis. *Advances in Methods and Practices in Psychological Science*, *4*(3). doi: 10.1177/25152459211031256

Heathcote, A., & Matzke, D. (2021). The limits of marginality. *Computational Brain & Behavior*, *6*(1), 28–34. doi: 10.1007/s42113-021-00120-3

Hinne, M., Gronau, Q. F., van den Bergh, D., & Wagenmakers, E.-J. (2020). A conceptual introduction to Bayesian model averaging. *Advances in Methods and Practices in Psychological Science*, *3*(2), 200–215. doi: 10.1177/2515245919898657

Hoekstra, R., Morey, R. D., Rouder, J. N., & Wagenmakers, E.-J. (2014). Robust misinterpretation of confidence intervals. *Psychonomic Bulletin & Review*, *21*(5), 1157–1164. doi: 10.3758/s13423-013-0572-3

Hoffmann, T., Hofman, A., & Wagenmakers, E.-J. (2022). A tutorial on Bayesian inference for the A/B test with R and JASP. *Methodology*, *18*, 239–277. doi: 10.5964/meth.9263

Holloway, I. D., & Ansari, D. (2009). Mapping numerical magnitudes onto symbols: The numerical distance effect and individual differences in children's mathematics achievement. *Journal of Experimental Child Psychology*, *103*(1), 17–29. doi: 10.1016/j.jecp.2008.04.001

Jeffreys, H. (1961). *Theory of probability*. Oxford: Oxford University Press.

Kanizsa, G. (1976). Subjective contours. *Scientific American*, *234*(4), 48–52. doi: 10.1038/scientificamerican0476-48

Kass, R. E., & Wasserman, L. (1995). A reference Bayesian test for nested hypotheses and its relationship to the schwarz criterion. *Journal of the American Statistical Association, 90*(431), 928. doi: 10.2307/2291327

Klugkist, I., Kato, B., & Hoijtink, H. (2005). Bayesian model selection using encompassing priors. *Statistica Neerlandica, 59*(1), 57–69. doi: 10.1111/j.1467-9574.2005.00279.x

Lambert, B. J. (2018). *A student's guide to Bayesian statistics.* Sage Publications.

Lee, M. D., & Wagenmakers, E.-J. (2014). *Bayesian cognitive modeling: A practical course.* Cambridge: Cambridge University Press.

Liang, F., Paulo, R., Molina, G., Clyde, M. A., & Berger, J. O. (2008). Mixtures of *g* priors for Bayesian variable selection. *Journal of the American Statistical Association, 103*(481), 410–423. doi: 10.1198/016214507000001337

Lindley, D. V., & Scott, W. F. (1984). *New Cambridge elementary statistical tables.* Cambridge: Cambridge University Press.

Ly, A., Marsman, M., & Wagenmakers, E. (2017). Analytic posteriors for Pearson's correlation coefficient. *Statistica Neerlandica, 72*(1), 4–13. doi: 10.1111/stan.12111

Ly, A., Raj, A., Marsman, M., Etz, A., & Wagenmakers, E.-J. (2018). Bayesian reanalyses from summary statistics: A guide for academic consumers. *Advances in Methods and Practices in Psychological Science, 3*, 367–374. doi: 10.1177/2515245918779348

Lyons, L. (2013). *Discovering the significance of 5 sigma.* Retrieved from https://arxiv.org/abs/1310.1284

Masson, M. E. J. (2011). A tutorial on a practical Bayesian alternative to null-hypothesis significance testing. *Behavior Research Methods, 43*(3), 679–690. doi: 10.3758/s13428-010-0049-5

McElreath, R. (2020). *Statistical rethinking: A Bayesian course with examples in R and Stan.* CRC Press.

Nelder, J. A. (2000). Functional marginality and response-surface fitting. *Journal of Applied Statistics, 27*(1), 109–112. doi: 10.1080/02664760021862

Rosenthal, R., & Gaito, J. (1963). The interpretation of levels of significance by psychological researchers. *The Journal of Psychology, 55*(1), 33–38. doi: 10.1080/00223980.1963.9916596

Rouder, J. N., Engelhardt, C. R., McCabe, S., & Morey, R. D. (2016). Model comparison in ANOVA. *Psychonomic Bulletin & Review, 23*(6), 1779–1786. doi: 10.3758/s13423-016-1026-5

Rouder, J. N., & Morey, R. D. (2018). Teaching Bayes' theorem: Strength of evidence as predictive accuracy. *The American Statistician, 73*(2), 186–190. doi: 10.1080/00031305.2017.1341334

Rouder, J. N., Morey, R. D., Speckman, P. L., & Province, J. M. (2012). Default Bayes factors for ANOVA designs. *Journal of Mathematical Psychology, 56*(5), 356–374. doi: 10.1016/j.jmp.2012.08.001

Rouder, J. N., Speckman, P. L., Sun, D., Morey, R. D., & Iverson, G. (2009). Bayesian *t* tests for accepting and rejecting the null hypothesis. *Psychonomic Bulletin & Review, 16*(2), 225–237. doi: 10.3758/pbr.16.2.225

Scott, J. G., & Berger, J. O. (2006). An exploration of aspects of Bayesian multiple testing. *Journal of Statistical Planning and Inference, 136*(7), 2144–2162. doi: 10.1016/j.jspi.2005.08.031

Scott, J. G., & Berger, J. O. (2010). Bayes and empirical-Bayes multiplicity adjustment in the variable-selection problem. *The Annals of Statistics, 38*(5). doi: 10.1214/10-aos792

van den Bergh, D., Clyde, M. A., Raj, A., de Jong, T., Gronau, Q. F., Marsman, M., . . . Wagenmakers, E.-J. (2021). A tutorial on Bayesian multimodel linear regression with BAS and JASP. *Behavior Research Methods, 53*, 2351–2371. doi: 10.3758/s13428-021-01552-2

van den Bergh, D., Wagenmakers, E.-J., & Aust, F. (2023). Bayesian repeated-measures ANOVA: An updated methodology implemented in JASP. *Advances in Methods and Practices in Psychological Science, 6*, 1–11. doi: 10.1177/25152459231168024

van Doorn, J., van den Bergh, D., Böhm, U., Dablander, F., Derks, K., Draws, T., . . . Wagenmakers, E.-J. (2021). The JASP guidelines for conducting and reporting a Bayesian analysis. *Psychonomic Bulletin & Review, 28*(3), 813–826. doi: 10.3758/s13423-020-01798-5

Wagenmakers, E.-J. (2007). A practical solution to the pervasive problems of p-values. *Psychonomic Bulletin & Review, 14*(5), 779–804. doi: 10.3758/BF03194105

Wagenmakers, E.-J., & Matzke, D. (2024). *Bayesian inference from the ground up: The theory of common sense.* Retrieved from https://bayesianspectacles.org/free-course-book

Wagenmakers, E.-J., Lodewyckx, T., Kuriyal, H., & Grasman, R. (2010). Bayesian hypothesis testing for psychologists: A tutorial on the Savage-Dickey method. *Cognitive Psychology, 60*(3), 158–189. doi: 10.1016/j.cogpsych.2009.12.001

Westfall, P., Johnson, W. O., & Utts, J. M. (1997). A Bayesian perspective on the Bonferroni adjustment. *Biometrika, 84*(2), 419–427. doi: 10.1093/biomet/84.2.419

Zellner, A., & Siow, A. (1980). Posterior odds ratios for selected regression hypotheses. In J. M. Bernardo, M. H. DeGroot, D. V. Lindley, & A. F. M. Smith (Eds.), *Bayesian statistics: Proceedings of the first international meeting* (pp. 585–603). Valencia: University of Valencia Press. doi: 10.1007/BF02888369

INDEX